Biotribology

Biotribology

Edited by
J. Paulo Davim

ISTE

WILEY

First published 2010 in Great Britain and the United States by ISTE Ltd and John Wiley & Sons, Inc.

ISTE Ltd
27-37 St George's Road
London SW19 4EU
UK

www.iste.co.uk

John Wiley & Sons, Inc.
111 River Street
Hoboken, NJ 07030
USA

www.wiley.com

Library of Congress Cataloging-in-Publication Data

Davim, J. Paulo.
Biotribology / J. Paulo Davim.
p. ; cm.
Includes bibliographical references and index.
ISBN 978-1-84821-275-6
1. Artificial joints. 2. Tribology. I. Title.
[DNLM: 1. Joint Prosthesis. 2. Biomechanics. 3. Surface Properties. WE 312]
RD686.D28 2010
617.5'80592--dc22

2010028015

British Library Cataloguing-in-Publication Data
A CIP record for this book is available from the British Library
ISBN 978-1-84821-275-6

Printed and bound in Great Britain by CPI Antony Rowe, Chippenham and Eastbourne.

MIX
Paper from responsible sources
FSC www.fsc.org FSC® C013604

Table of Contents

Preface

Tribology is the "*science and technology of interacting surfaces in relative motion*" and embraces the study of friction, wear, and lubrication. By extension, *biotribology* is usually defined as the tribological phenomena occurring in either the human body or in animals. Therefore it is possible to consider tribological processes that may occur after implantation of an artificial device in the human body and the tribological processes naturally occurring in or on the tissues and organs of animals. Animals, including humans, possess a wide variety of sliding and frictional interfaces.

The purpose of this book is to present a collection of examples illustrating the state-of-the-art and research developments in biotribology. Chapter 1 provides the biotribology of total hip replacement, the metal-on-metal articulation. Chapter 2 contains experimental wear studies of total joint replacements. Chapter 3 covers the influence of temperature on creep and deformation in UHMWPE under tribological loading in artificial joints. Chapter 4 contains information on large capacity wear testing. Finally, Chapter 5 is dedicated to biotribology of titanium alloys.

This book can be used as a textbook for a final undergraduate engineering course or as a reference on biotribology for postgraduate level. This book can also serve as a useful reference for academics, materials and biomedical researchers, materials, mechanical and biomedical engineers, professionals in biotribology and medical tribology and related industries. The scientific relevance of this book is evident for many important research centers, laboratories and universities throughout the world. Therefore, it is hoped that this book will encourage and enthuse other research in this recent field of science and technology.

I would like to acknowledge my gratitude to ISTE-Wiley for this opportunity and for their professional support. Finally, I would like to thank all the authors of the chapters for their input and their availability for this work.

J. Paulo DAVIM
University of Aveiro, Portugal
July 2010

Chapter 1

Biotribology of Total Hip Replacement: Metal-on-Metal Articulation

1.1. Introduction

All structures of the human body undergo a natural process of aging over the course of life. This includes articular cartilage. Extreme wear of articular cartilage of the hip joint can be disabling and require treatment. The reconstruction of joint function by means of a hip joint replacement (see Figure 1.1) has been described as one of the most successful medical procedures [BER 05]. Most patients regain functionality of their affected joint within a brief period after a surgical intervention. Worldwide, about one million artificial hip joints are implanted annually.

The bearing surfaces of implants are subjected to friction and wear. Particles produced in the contact area between bearing partners are released into the tissue surrounding the implant. These particles can provoke inflammatory

---

Chapter written by J. Philippe KRETZER.

tissue reactions, which can result in bone loss. Massive bone loss around an implant jeopardizes the fixation of the implant and results in loosening. This process, referred to as aseptic loosening, necessitates surgical revision of the implant. The frequency of revisions has increased dramatically in recent years.

Currently, one out of five artificial hips is subjected to revision within 15 years [KÄR 08]. Revision surgeries are complicated, risky and expensive, and revised implants are generally less successful than primary implants.

The recent rise in the use of joint replacements in young and active patients demands a drastic increase in implant durability. The selection of suitable materials plays an important role in the pursuit of this goal.

Currently, three main classes of materials are being used in hip replacements: ceramics, metals and polymers. The ceramics category includes aluminum oxide ceramics and zirconia-toughened alumina ceramics. The polymer category is dominated by ultra-high molecular weight polyethylene (PE), both in its conventional and cross-linked (X-PE) forms. Cobalt-chrome-molybdenum alloys (CoCrMo) make up the metal category.

These materials can be combined in several ways and are generally classified in hard-soft and hard-hard combinations. Hard-soft combinations consist of a metal or ceramic head paired with a PE or X-PE cup. Hard-hard combinations pair up two ceramic components (ceramic-on-ceramic) or two metal components (metal-on-metal). Hard-soft combinations are currently the most common. The prevalence of metal-on-metal bearings is however rising, in part due to their popularity among young and active patients.

1.2. Historical development of metal-on-metal bearings in total hip replacements

Total hip replacements took off around the world after the introduction of PE as a bearing material by Sir John Charnley in the 1960s [CHA 61]. While Charnley was developing PE on metal bearings, others were experimenting with metal-on-metal combinations. Most early developments took place in Great Britain. The McKee-Farrar prosthesis was implanted for the first time in 1950, followed by the Stanmore prosthesis in 1963 and the Ring prosthesis one year later [AMS 96]. The McKee-Farrar prosthesis (see Figure 1.1) was the most widely used metal-on-metal bearing in the 1960s and early 1970s [AMS 01]. The early success of these metal-on-metal bearings was, however, limited by a high and unacceptable failure rate [VIS 87]. Promising results of Charnley's PE bearing and the problems surrounding metal-on-metal bearings caused the latter to virtually disappear. McKee-Farrar ceased to implant his device after 1972 and PE bearings became the gold standard [AMS 96].

Aseptic loosening, which has been associated with PE particles, and the comparatively low wear of metal-on-metal bearings encouraged Müller and Weber in 1988 to bring a second generation of metal-on-metal bearings to the market (see Figure 1.1) [WEB 96]. The Swiss company Sulzer (now Zimmer) offers this bearing under the name Metasul® to a worldwide market. To date, more than 200,000 Metasul® devices have been implanted [TIP 05].

Since the introduction of the second generation, metal-on-metal bearings have been on the rise. Over the past few years, many developments have contributed to continuous improvement and third generation implants are now in use. An example of recent advances is metal-on-metal hip surface replacements (see Figure 1.1).

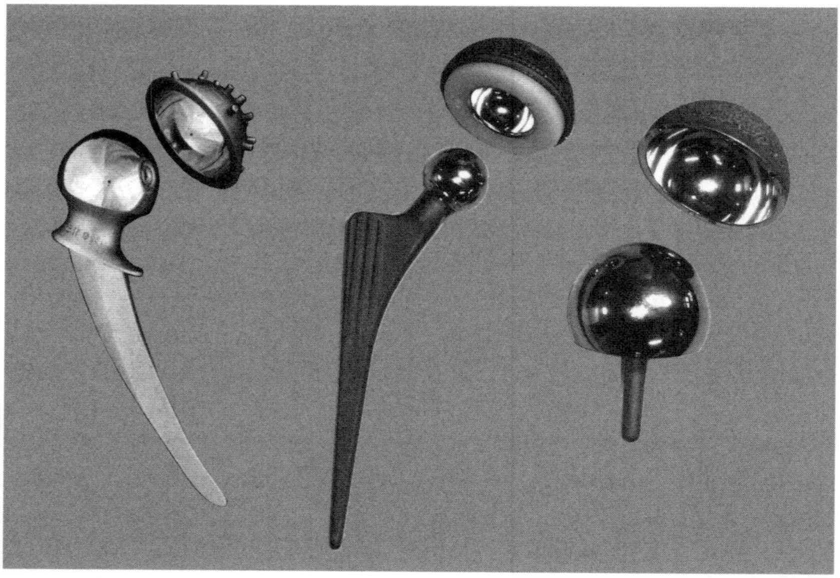

Figure 1.1. *Metal-on-metal bearings in total hip replacements. Left to right: McKee-Farrar, Metasul, BHR surface replacement*

1.3. Design and materials

1.3.1. *Implant geometry*

Figure 1.2 is a schematic cut through a metal-on-metal surface replacement. Ideally, head and cup are two ball-shaped elements with slightly differing diameters. The small space between head and cup is referred to as radial clearance. No surface, no matter how well polished, is completely smooth and flat. The head and cup therefore deviate slightly from perfect spheres, resulting in deviations of roundness. The roughness of an implant surface is specified by the mean roughness, calculated as the arithmetic mean of absolute values. Geometric indicators used to characterize bearing partners include diameter, clearance, deviation of roundness and surface roughness.

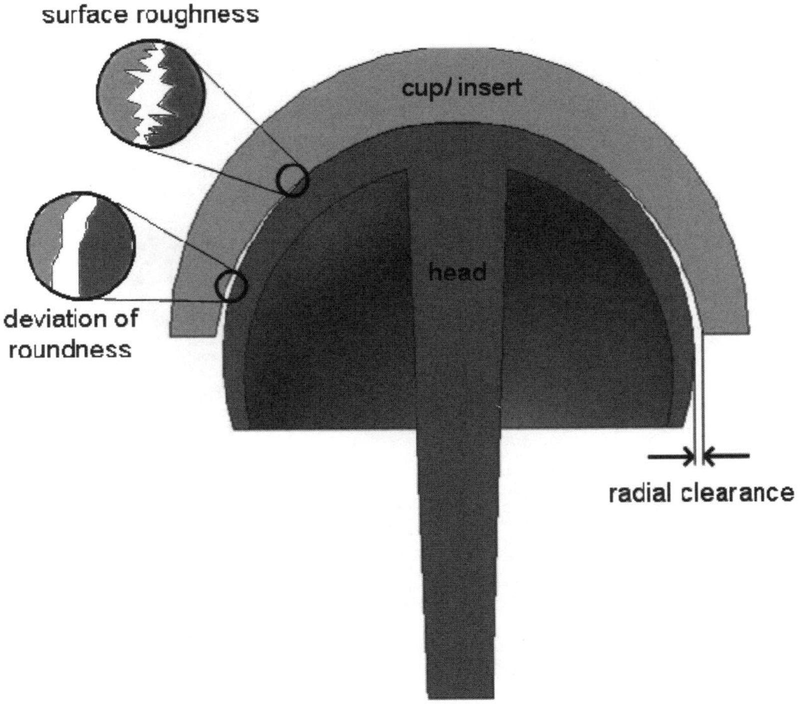

Figure 1.2. *Schematic of implant geometry*

1.3.2. *Manufacturing methods and metallurgy*

$Co_{28}Cr_6Mo$ is the most widely used alloy in metal-on-metal bearing material due to its good wear properties and high corrosion resistance. Traces of nickel, manganese or iron are also sometimes included. Two chemically virtually identical – but in their mechanical properties different – cobalt alloys are in use. Based on their manufacturing methods, these alloys are classified as wrought (ISO 5832-12, ASTM F1537) and cast (ISO 5832-4, ASTM F75) materials. In addition to the manufacturing method, carbon content critically influences metallurgical properties and microstructure. In this context, materials are classified as low carbon (lc, $< 0.15\%$) and high carbon (hc, $\geq 0.15\%$).

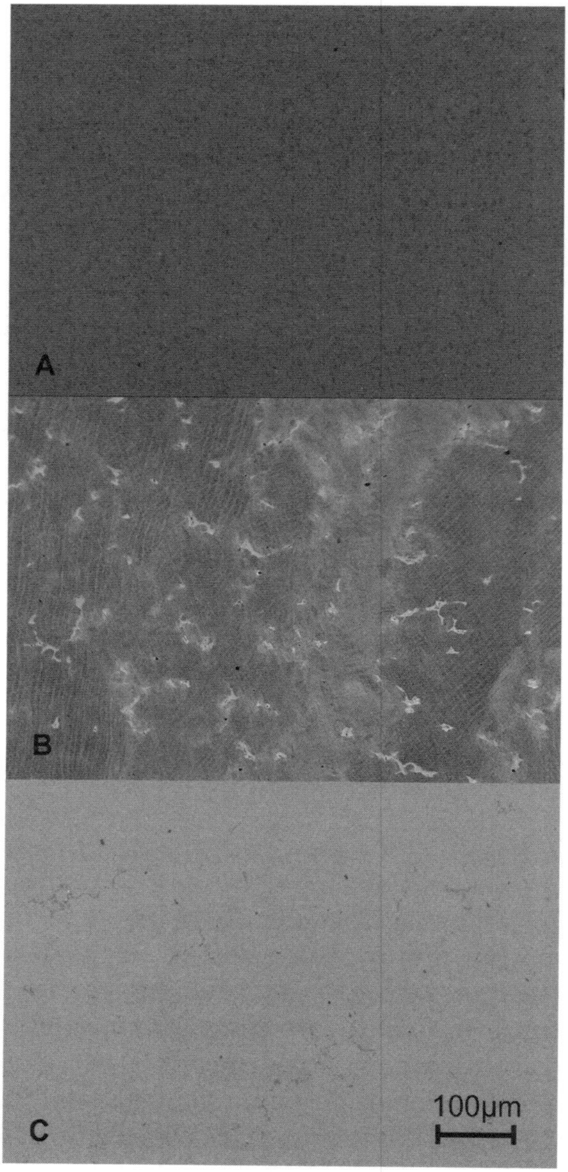

Figure 1.3. *SEM Images of different CoCrMo Alloys: A) microstructure of a wrought CoCrMo alloy with high carbon content; B) microstructure of a cast CoCrMo alloy with high carbon content; C) microstructure of a cast CoCrMo alloy with high carbon content after heat treatment (HIP, SA)*

1.3.2.1. *CoCrMo wrought alloys (ISO 5832-12, ASTM F1537)*

This alloy is used to manufacture hip replacement components through mechanical treatment or serves as a raw material for thermomechanical forming.

The microstructure is comprised of approximately 15-20 µm sized grains and at low carbon contents is monophasic and purely austenitic (face-centered cubic, fcc) [STR 96]. Increasing carbon content gives rise to chromium and molybdenum carbides, which results in a biphasic structure [LEE 06] (see Figure 1.3-A).

Carbides of the type $M_{23}C_6$ and M_7C_3 are about five times harder than an austenitic matrix [STR 96, LEE 06, SCH 96]. They are relatively small (approx. 3 µm) and regularly dispersed [STR 96, TIP 05]. These carbides cover about 5% of the surface area.

Under loading, a stress-induced face centered cubic (fcc) to hexagonal close-packed (hcp) martensitic transformation is possible. Martensite is harder and has impressive tribological properties [VAR 03, SCH 96]. Rising carbon content leads to higher tensile strength and ductility [STR 96, LEE 06]. The mechanical properties of wrought alloys are roughly twice as high as the properties of cast alloys (Table 1.1) [STR 96].

The ISO and ASTM standards differ only marginally in their mechanical demands. The chemical alloy definitions are the same in the two standards. However, the ASTM standard deviates slightly from the literature [TIP 05] in its classification of carbon content and defines low carbon in the range of 0-0.14% and high carbon in the range of 0.15-0.35%.

	Cast high carbon	Wrought high carbon	Wrought low carbon
Carbon content in %	>0.20	>0.20	<0.08
Grain size (average) in µm	2000	20	15
Carbide size (average) in µm	30	3	-
Hardness HV	420	470	430
0.2% - Yield strength in MPa	450-600	830-1,200	500-950
Tensile strength in MPa	650-1,000	1,170-1,600	1,000-1,374
Bending fatigue strength in MPa	190-400	500-870	450-830

Table 1.1. *Comparison of cobalt alloys according to [STR 96] (© 1996, with permission from Professional Engineering Publishing)*

1.3.2.2. *CoCrMo cast alloys (ISO 5832-4, ASTM F75)*

Cast implants are manufactured using the investment casting method, also referred to as lost wax casting. This process involves the creation of a master pattern from wax, which is subsequently coated with a ceramic material. The ceramic layer can be used as a mould upon melting and removal of the wax. This method makes the production of complex geometries such as knee and hip implants possible [MET 04].

As is the case for wrought alloys, the ISO and ASTM standards for cast alloys are only marginally different with regard to the chemical composition of the alloys. The mechanical requirements are the same in both standards.

In cast alloys the number and size of carbides rises with increasing carbon content. With a grain size of 2,000 µm, the microstructure is much coarser than the structure of a wrought alloy. The size of dendritic blocky carbides in fcc cobalt-rich matrices is about one order of magnitude larger than the carbides in wrought alloys [STR 96, SHI 93].

The structure of cast components often exhibits pores or holes, resulting in reduced mechanical properties [CAW 03]. Some manufacturers use hot isostatic pressing (HIP) and solution annealing (SA) in order to close these porosities and to achieve a homogeneous microstructure [NEV 04].

Non-heat treated, cast implant components are referred to by the term "as cast". The HIP procedure exposes components to a temperature of 1,200°C and a pressure of 103 MPa in an argon atmosphere for about four hours. The primary goal of the HIP procedure is the closure of internal porosities [MET 04, NEV 04].

The SA procedure exposes components to a temperature of 1,200°C in a vacuum, followed by cooling in nitrogen. This process improves ductility and leads to a homogeneous microstructure [DOB 83, NEV 04]. Heat treatment causes blocky M_7C_3 type carbides to dissolve (see Figure 1.3-B) without a phase change in the matrix, leading to lamellar and insular $M_{23}C_6$ type carbides (see Figure 1.3-C) [CAU 02, VAR 03]. The carbide content is markedly reduced by heat treatment [CAW 03, MET 04].

In some surface replacements the cup is manufactured with a porous coating, which is supposed to facilitate bone ingrowth. This coating is applied in a sintering process similar to the HIP and SA heat treatments.

The main manufacturing- and material-dependent criteria influencing the mechanical and metallurgical properties of hip implant bearings include: carbon content of the alloy, primary manufacturing method (cast or wrought) and heat treatment.

1.4. Tribology of metal-on-metal bearings in total hip replacement

1.4.1. *Wear and types of friction*

Lubrication is a key factor defining wear and frictional properties of a hip implant. Friction and wear can be greatly reduced by a lubricating film separating the bearing partners. The mode of friction varies depending on implant geometry, surface roughness, material elasticity, load on and relative velocity of the bearing partners, as well as viscosity of the lubricant.

The Stribeck curve (see Figure 1.4) shows different modes of friction. Coefficient of friction (μ) and film thickness (h) are shown as a function of the Sommerfeld number, which is unitless and based on viscosity (η), velocity (v) and pressure (p).

Given a narrowing gap, the sequence of modes of friction and increasing film thickness is as follows:

– boundary lubrication;

– mixed film lubrication;

– elastohydrodynamic lubrication;

– hydrodynamic lubrication.

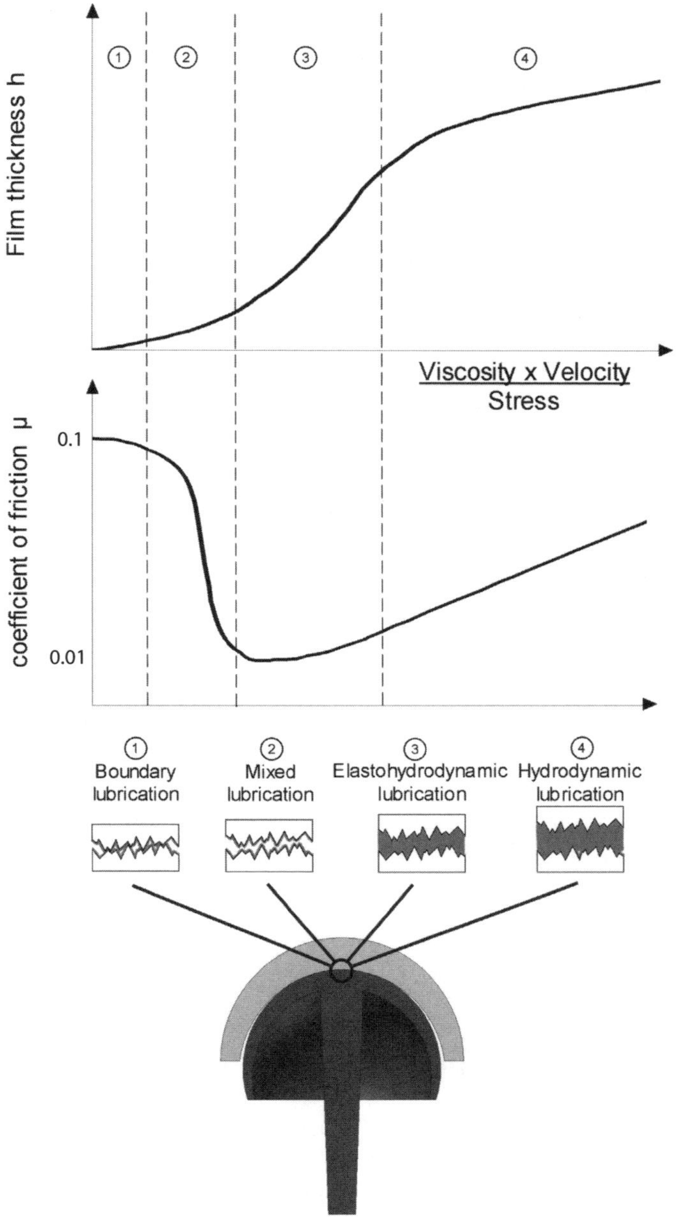

Figure 1.4. *Stribeck curve*

For boundary lubrication, the surface of the bearing partners is coated with a thin film of lubricant molecules. The load is carried by asperities, while shear is transmitted via lubricant molecules. Friction and wear are determined primarily by the properties of the contact surfaces and secondarily by the properties of the lubricant.

Mixed film lubrication occurs if the gap between contact surfaces is partially but not fully covered by a lubricating film. The load is carried by the lubricating film and asperity forces alike. The properties of the contact surfaces and the lubricant both contribute to the frictional and wear properties.

Elastohydrodynamic lubrication (EHL) is characterized by a continuous, uninterrupted lubricating film between the bearing partners. EHL arises from elastic deformations of the components and takes into account a possible increase in viscosity of the lubricant under pressure. The dynamic viscosity of the lubricant alone governs the frictional properties since the bearing partners are not in contact. The bearing partners do not exhibit wear.

Hydrodynamic lubrication is characterized by a relatively thick, uninterrupted lubricating film. The bearing partners do not wear at all. Elastic deformations are not taken into account, and rigid body geometries as well as constant viscosity of lubricants are assumed [CZI 03].

1.4.2. *EHL theory of lubrication*

The pressure distribution in the lubricating film is based on Reynolds' equation [HAM 78]. The pressure in the lubricating film must be high enough to separate the bearing partners. This pressure depends on the geometry of the contact area, the moduli of elasticity of the bearing partners, the relative velocity and properties of the lubricant. The minimal lubricating film thickness for metal-on-metal hip implant bearings can be calculated based on the Hertz theory [JIN 97]:

$$h_{\min} = R' \cdot 2.8 \cdot \left(\frac{\eta \cdot u}{E' \cdot R}\right)^{0.65} \cdot \left(\frac{F}{E' \cdot R^2}\right)^{-0.21} \quad [1.1]$$

where:

h_{min}	min. film thickness	in µm
R’	equivalent radius of curvature	in mm
η	viscosity	in Pa·s
u	entry speed	in m/s
E’	equivalent elasticity	in N/mm^2
F	contact force	in N

The equivalent radius of curvature *R’* describes the clearance geometry based on the difference between head and cup diameter. It is equivalent to the diameter of a sphere gliding on a flat surface and is calculated as follows [JIN 97, JAL 06, JAL 01]:

$$R' = \frac{R_1 \cdot R_2}{R_2 - R_1} \quad [1.2]$$

where:

R_1	head radius	in mm
R_2	cup radius	in mm

The equivalent elasticity *E'* is used to describe elastic deformations at the contact surface. *E'* characterizes the overall elasticity of head and cup and is comprised of the moduli of elasticity and Poisson's ratios of the materials [JIN 97, MED 01]:

$$E' = 2\left(\frac{1-\nu_1^{\ 2}}{E_1} + \frac{1-\nu_2^{\ 2}}{E_2}\right)^{-1} \qquad [1.3]$$

where:

ν_1	Poisson's ratio head	dimensionless
ν_2	Poisson's ratio cup	dimensionless
E_1	modulus of elasticity head	in N/mm^2
E_2	modulus of elasticity cup	in N/mm^2

The surface roughness of both components plays an important role in describing the mode of friction to guarantee separation of the contact surfaces. The lower the surface roughness, the thinner the minimal lubrication film thickness needs to be in order to separate the bearing partners.

No surface, no matter how well polished, is completely smooth and flat. The mean roughness index R_a is used to classify the mode of friction of the bearing couple. R_a is defined as the arithmetic mean deviation of the roughness profile.

The ratio of surface roughness to minimal lubricating film thickness is specified by the dimensionless constant lambda and is calculated as follows [DOW 06, MED 01]:

$$\lambda = \frac{h_{\min}}{\sqrt{\left((R_{a1})^2 + (R_{a2})^2\right)}} \quad [1.4]$$

where:

λ	lambda ratio	dimensionless
R_{a1}	mean roughness index head	in μm
R_{a2}	mean roughness index cup	in μm

If a value of $\lambda > 3$ is reached, the lubricating film is thick enough to separate the bearing surfaces completely (elastohydrodynamic lubrication). If $\lambda < 1$, the lubricating film is too thin to separate the surfaces completely, resulting in boundary lubrication. Mixed film lubrication exists for λ between 1 and 3 [DOW 06, MED 01].

Figure 1.5 depicts the EHL mode of lubrication and exemplifies the influence of the clearance between head and cup and the implant diameter on the lubricating film thickness [JIN 97].

For calculations, bearing diameters of 28 mm and 54 mm were chosen. A CoCrMo alloy with a modulus of elasticity of $E_1 = E_2 = 230{,}000$ N/mm^2 and a Poisson's ratio of $\nu_1 = \nu_2 = 0.3$ was assumed. A surface roughness of $R_{a1} = R_{a2} = 0.02$ μm [HEI 09], a contact force of $F = 2{,}500$ N [BER 93], and lubricant viscosity of $\eta = 0.005$ Pa·s [JIN 97] were assumed.

The entry speed u is equivalent to one half of the relative velocity between head and cup. For a diameter of 28 mm, the relative velocity during regular walking is 0.02 m/s. The relative velocity is about twice as high for a diameter of 54 mm.

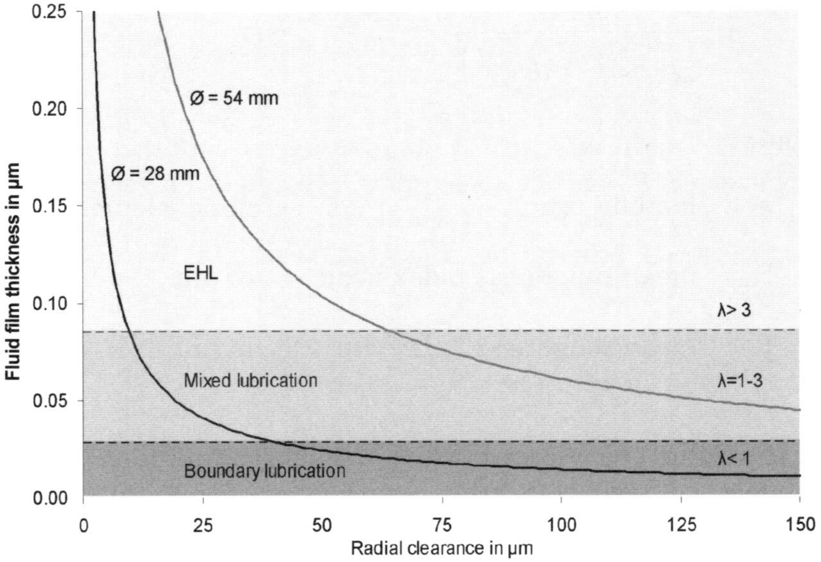

Figure 1.5. *Lubricating film thickness as a function of clearance*

Figure 1.5 shows the minimal lubricating film thickness as a function of clearance depending on two different diameters. EHL theory reveals the direct influence of head diameter and clearance on lubricating film thickness. A larger head diameter and smaller clearance result in a thicker lubricating film.

A surface roughness of 0.02 µm and a clearance of 50 µm lead to boundary lubrication ($\lambda = 1$, $h_{min} = 0.028$ µm) for a 28 mm implant diameter, whereas implants 54 mm in diameter would lead to EHL ($\lambda = 3$, $h_{min} = 0.085$ µm).

This model is severely limited by the assumption of constant relative velocities. This does not correspond to *in vivo* conditions, which include relative velocities close to zero at the transition from flexion to extension. EHL also does not take into account deviations of roundness of implant components. These deviations are small (<10 μm for modern surface replacements) but cannot be ignored [HEI 09].

Temperature-dependent effects and varying wettability of different materials are not considered in this model. However, EHL theory does show the effect of geometrical variables such as implant diameter, clearance and surface roughness on lubricating film thickness and the resulting impact on wear properties.

1.4.3. *Friction in physiological joints*

Friction in natural joints is described as EHL [UNS 91, JIN 93, WIL 96]. This mode of lubrication is possible due to the elastic deformations of cartilage. The modulus of elasticity of cartilage is much smaller than the modulus of elasticity of materials such as PE, CoCrMo or ceramics.

Locally acting pressure decreases the surface roughness of cartilage from about 0.07-0.45 μm when unloaded to about 0.01-0.03 μm under loading [DOW 06, MOA 03, CLA 99]. The coefficient of friction of natural joints is about $\mu = 0.005$-0.037 [CHA 60, CAL 07, ANT 07] and is very low compared to the coefficient of friction of CoCrMo on CoCrMo ($\mu = 0.22$-0.27) or CoCrMo on PE ($\mu = 0.06$-0.08) [SCH 00].

1.4.4. *Friction in artificial joints*

The mode of friction in artificial joints depends on the materials used as bearing partners. Hard-soft bearings (PE-on-metal, PE-on-ceramic) exhibit boundary lubrication [UNS 91, DOW 06, WIL 96, SAI 97, SCH 00, WID 02]. The literature cites the high surface roughness of PE as a reason for this choice [JIN 97].

This explanation is, however, not sufficient since the surface roughness of PE changes due to cold flow. For example, PE surfaces of surgically retrieved hip cups can have very smooth surfaces.

Friction of polymers is caused by deformation and adhesion (van der Waals forces). The adhesive component of friction is small if, for example, a hard sphere rolls on a polymeric surface. Friction mainly arises from deformation energy at the contact surface and results from the viscoelasticity of the polymer [BRI 98]. Boundary lubrication of hard-soft bearings in joint replacement is based on the viscoelasticity of the polymer [ZHA 98, WID 02].

The mode of friction of hard-hard bearings (metal-on-metal, ceramic-on-ceramic) has been investigated experimentally. Friction for small diameter implants (16-28 mm) has been characterized as boundary lubrication, whereas larger diameter implants (36 mm/54 mm) have been shown to exhibit mixed film lubrication and sometimes EHL [DOW 04a, DOW 00, BRO 07].

It has also been shown that the coefficient of friction decreases during the wear process. Boundary lubrication can progress to mixed film lubrication and even elastohydrodynamic lubrication [VAS 06].

Experimental evidence indicates that mixed film lubrication dominates in metal-on-metal bearings. Piecewise EHL is possible during the swing phase or with larger head diameters.

1.5. Wear testing

1.5.1. *Simulation in hip simulators*

Tribological investigations are used to determine the wear properties of implants prior to clinical trials and to allow for analysis of debris. This can be done using hip simulators (see Figure 1.6).

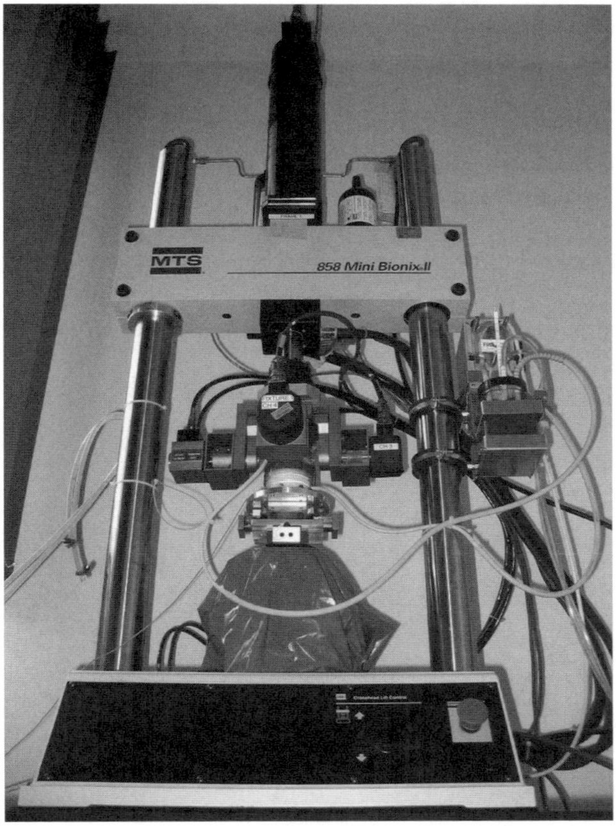

Figure 1.6. *Hip simulator for wear testing on implants*

Simulators should expose implant components to physiological movement patterns and loads as well as mimic realistic physiological environments as closely as possible.

Methodological variations between simulator studies carried out in different labs around the world are significant and the discussion of "optimal" simulation parameters is ongoing.

Simulators differ in their degrees of freedom, kinematics and loads applied to implants. Furthermore, there is neither consensus on implant positioning nor composition and temperature of test mediums.

The International Organization for Standardization (ISO) has published guidelines with the goal of standardizing wear tests on hip implants.

ISO 14241-1:2002(E) is based on the natural gait cycle of humans for walking on level surfaces. Movement of the femur relative to the pelvis is simulated using three degrees of freedom (see Figure 1.7).

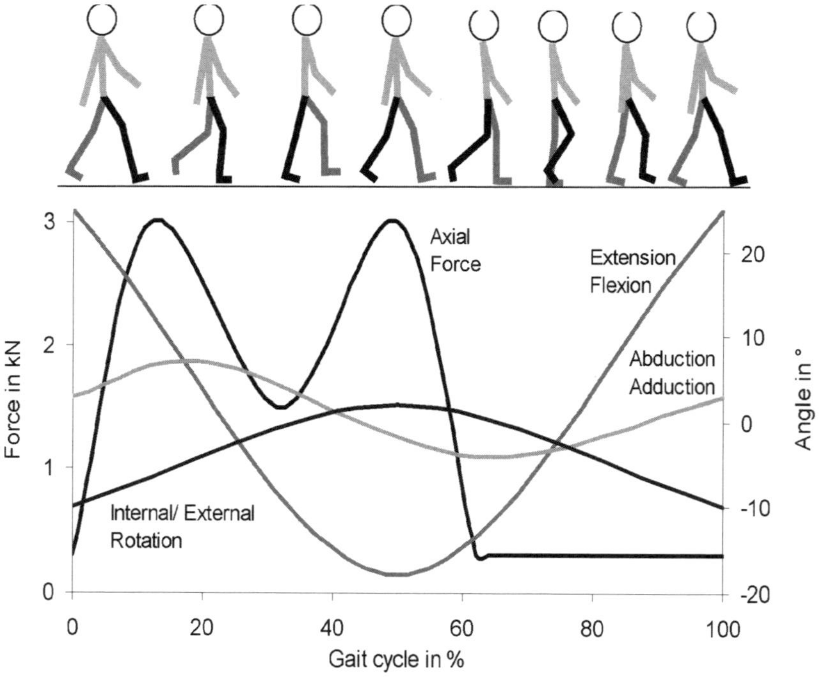

Figure 1.7. *Gait cycle for wear analysis of hip implants based on ISO 14241-1:2002(E)*

The movement pattern depicted (see Figure 1.7) includes flexion/extension (FL/EX) of +25°/0/-18°, ab-/ adduction

(AB/AD) of -4°/0/+7° and internal/external rotation (IR/ER) of -10°/0/+2°. The resulting hip contact force (F) upon heel strike is 3 kN. The force drops to 1.5 kN during the stance phase, rises to 3 kN during toe off, and drops to 300 N during the swing phase.

The ISO standard requires the use of calf serum at 37°C as a test medium. The prescribed 5 million cycles at a frequency of 1 Hz allows for simulation of 3-5 years *in vivo* during the course of a 10-week test.

Recently, an additional standard, ISO 14242-3:2009(E), for wear testing using orbital type simulators was published.

1.5.2. *Wear determination*

Wear is measured during the course of the simulation. This is usually achieved by geometric or gravimetric measurements. The ISO 14241-2:2000(E) standard specifies parameters for wear measurement.

In the geometric method, the surfaces of the articulating components are analyzed based on a tight network of measured points using a coordinate measurement machine. Volumetric differences representing the wear volume are calculated based on the resulting clouds of data points (see Figure 1.8). Alternatively, roundness can be measured.

The geometric method is not commonly used in simulator studies since the very low wear of hard-on-hard bearings (e.g. metal-on-metal, ceramic-on-ceramic) would necessitate extremely high measurement accuracy and hence very long measurement times. Coordinate measurement is, however, suitable for the analysis of retrieved implants.

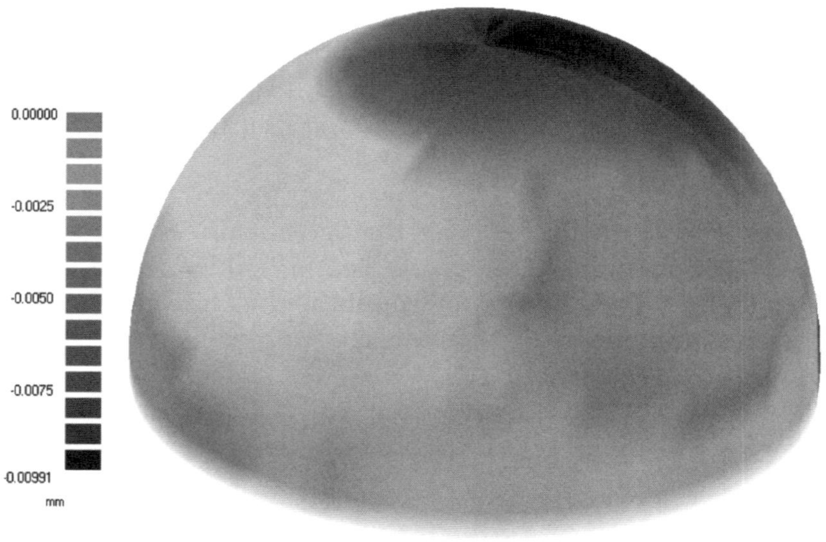

Figure 1.8. *Geometric wear analysis based on coordinate measurement on the head*

The most common method used to determine wear in simulator studies is gravimetric wear measurement. The change in weight of a component during a simulation as determined in pre- and post-simulation measurements is used to calculate the wear volume using component density.

An alternative method of wear measurement, which is especially useful for metal-on-metal bearings, is ion analysis. In this method, the ion concentration in the test medium is measured after digestion of the medium. Many challenges of gravimetric wear measurement can be avoided when using this method.

Wear analysis can be carried out frequently without removal of components from the simulator. The method is extremely precise and yields detailed information about the chemical composition of the wear products. However, ion analysis requires a contaminant-free simulation environment, a clean room for sample processing and suitable analytical devices [KRE 10].

1.5.3. *Wear properties*

The characteristic wear properties of a metal-on-metal bearing are depicted in Figure 1.9. Metal-on-metal bearings typically exhibit running-in wear, a period during which the two bearing surfaces improve their fit. This phase is characterized by an elevated wear rate. Following the running-in phase, the wear volume is reduced to the rather constant steady-state wear rate.

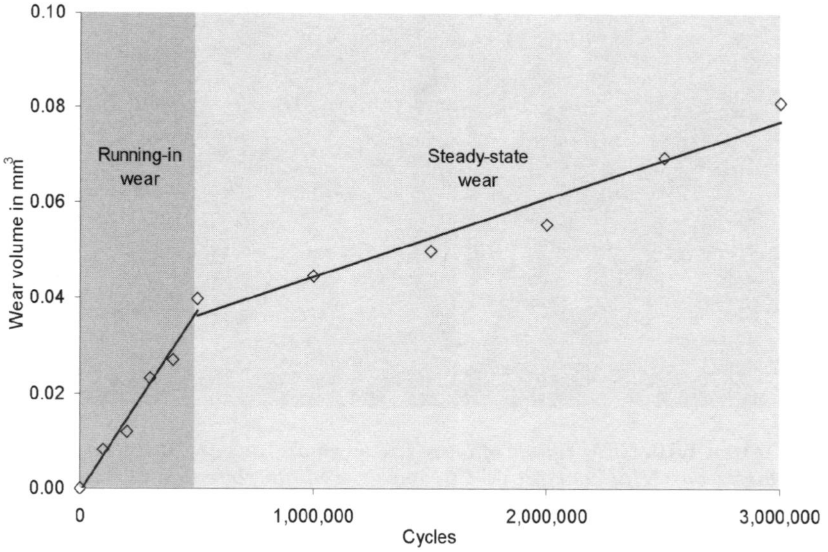

Figure 1.9. *Wear properties of a metal-on-metal bearing*

Running-in wear rate, steady-state wear rate, length of the running-in phase and total wear volume (dependent on simulation duration) are established wear characteristics.

Metal-on-metal bearings exhibit both mechanical and chemical wear mechanisms. Abrasion is dominated by micro-chipping and microscopic propagation of grooves and scratches (see Figure 1.10-C). In the case of adhesion, chemical processes (tribochemical reactions) are a primary factor. Molecular interactions on the boundary surface cause adhesion of the bearing partners. Boundary surface fragments are pulled out of the surface due to movement of the bearing partners, resulting in holes and craters (see Figure 1.10-B).

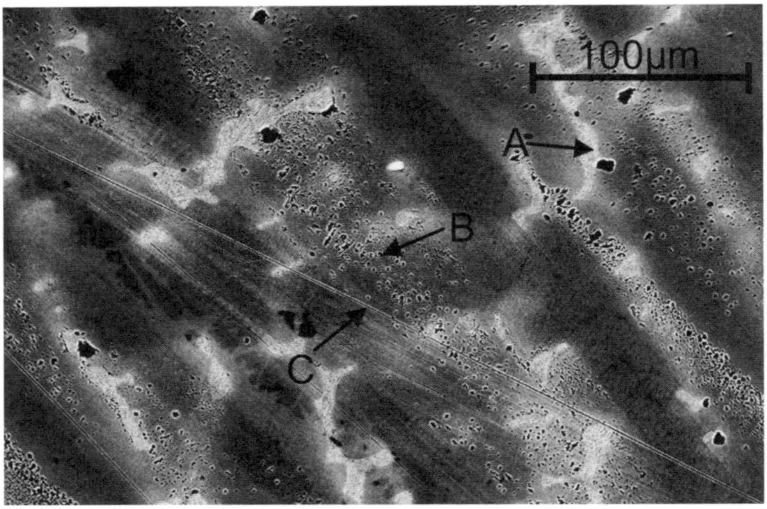

Figure 1.10. *REM image of wear traces on an "as cast", high carbon metal-on-metal bearing. A) Chromium and molybdenum carbides, B) traces of adhesive wear, C) abrasive wear*

1.5.4. *Results of wear tests*

Table 1.2 summarizes the results of numerous simulator studies on metal-on-metal implants [KRE 09]. This overview solely includes studies indicating not only the amount of wear (by gravimetric method) but also information about implant geometry (diameter and clearance), manufacturing method (as cast, wrought, heat treated) and carbon content.

#	diameter mm	radial clearance µm	carbon content				manufacturing-heat treatment						RI wear rate ± SD mm³/10⁶ cyc	SS wear rate ± SD mm³/10⁶ cyc	RI/SS ratio	n	author	year
			head		cup		head			cup								
			hc	lc	hc	lc	wr	ac	HT	wr	ac	HT						
1	28	22.0		x		x	x			x			0.69 *	0.10 *	6.9	2	SCH 01	2001
2	28	22.5		x		x	x			x			1.37 *	0.32 ± 0.01	4.3	2	FIR 01	2001
3	28	28.0	x		x		x			x			0.32 *	0.02 < 0.00	13.9	2	FIR 01	2001
4	28	28.0	x		x			x			x		0.21 ± 0.14	0.06 ± 0.05	3.3	8	CHA 99	1999
5	28	30.0		x	x		x			x			3.00 ± 0.40	0.80 ± 0.08	3.8	3	FIS 04	2004
6	28	30.0		x	x		x			x			2.20 ± 0.18	0.30 ± 0.10	7.3	3	FIS 02	2002
7	28	30.5		x	x		x			x			1.60 *	0.54 *	3.0	3	SMI 01	2001
8	28	31.3	x		x		x			x			2.25 *	1.10 *	2.0	4	DOW 04a	2004
9	28	32.5	x		x		x			x			0.24 ± 0.22	0.07 ± 0.02	3.6	6	CHA 99	1999
10	28	32.8	x		x		x			x			0.85 ± 0.24	0.03 ± 0.02	28.3	4	LIA 07	2007
11	28	33.3	x			x	x			x			0.30 *	0.04 < 0.00	8.1	2	FIR 01	2001
12	28	40.0		x		x	x			x			0.82 *	0.25 *	3.3	2	SCH 01	2001
13	28	44.9	x		x		x			x			0.25 ± 1.25	0.11 ± 0.03	2.3	5	JOH 99	1999
14	28	42.7		x		x	x			x			0.25 ± 0.23	0.15 ± 0.01	1.7	5	JOH 99	1999

Table 1.2. a) *Wear of metal-on-metal bearings according to [KRE 09]*
(© 2009, with permission from John Wiley & Sons)

#	diameter mm	radial clearance µm	carbon content head		carbon content cup		manufacturing-heat treatment head			manufacturing-heat treatment cup			RI wear rate ± SD mm³/10⁶ cyc	SS wear rate ± SD mm³/10⁶ cyc	RI/SS ratio	n	author	year
			hc	lc	hc	lc	wr	ac	HT	wr	ac	HT						
15	28	45.0	x		x			x			x		1.67 ± 0.14	0.54 < 0.00	3.1	4	AFF 06	2006
16	28	47.2	x		x		x			x			0.43 ± 0.26	0.08 ± 0.03	5.4	4	MED 98	1998
17	28	49.3	x		x		x			x			2.22 ± 0.72	0.98 ± 0.82	2.3	6	ANI 01	2001
18	28	50.0	x		x		x			x			0.31 ± 0.18	0.04 ± 0.01	7.8	4	LIA 04	2004
19	28	50.0		x		x	x			x			0.76 ± 0.51	0.11 ± 0.06	6.9	8	CHA 99	1999
20	28	53.5		x		x	x			x			0.60 ± 0.14	0.06 < 0.00	10.0	2	MED 98	1998
21	28	56.3	x		x		x			x			*	0.45 ± 0.27		4	GOL 00	2000
22	28	70.0		x	x		x			x			0.82 ± 0.31	0.06 ± 0.01	13.7	4	LIA 04	2004
23	28	110.0	x		x				x			x	4.20 ± 0.90	0.90 ± 0.10	4.7	4	BOW 04	2004
24	36	52.5	x		x		x			x			2.32 ± 0.02	*		3	DOW 04a	2004
25	36	52.5	x		x			x			x		0.76 ± 0.29	0.35 ± 0.26	2.2	4	AFF 06	2006
26	36	52.5	x		x			x				x	2.66 ± 0.08	*		3	DOW 04b	2004
27	36	52.5	x		x		x			x			3.36 ± 0.42	*		3	DOW 04b	2004

Table 1.2. b) *Wear of metal-on-metal bearings according to [KRE 09]*
(© 2009, with permission from John Wiley & Sons)

#	diameter mm	radial clearance µm	carbon content head hc	carbon content head lc	carbon content cup hc	carbon content cup lc	manufacturing-heat treatment head wr	manufacturing-heat treatment head ac	manufacturing-heat treatment head HT	manufacturing-heat treatment cup wr	manufacturing-heat treatment cup ac	manufacturing-heat treatment cup HT	RI wear rate ± SD mm³/10⁶ cyc	SS wear rate ± SD mm³/10⁶ cyc	RI/SS ratio	n	author	year
28	36	59.5	x		x			x			x		2.32 ± 0.58	*		4	DOW 04b	2004
29	36	60.5	x		x			x			x		3.30 ± 0.10	0.38 ± 0.04	8.7	3	HAR 06	2006
30	36	61.5	x		x			x			x		3.65 ± 0.67	*		5	DOW 04b	2004
31	36	61.8	x		x		x			x			2.80 ± 0.10	0.42 ± 0.04	6.7	2	HAR 06	2006
32	36	62.0	x		x		x			x			3.69 ± 0.14	*		2	DOW 04b	2004
33	36	63.0	x		x			x				x	3.10 ± 0.52	*		3	DOW 04b	2004
34	36	71.0	x		x		x			x			*	0.36 ± 0.62		6	GOL 00	2000
35	36	71.5	x		x		x			x			3.76 ± 0.43	*		4	DOW 04b	2004
36	36	72.0	x		x		x			x			3.51 ± 0.15	*		4	DOW 04a	2004
37	36	73.5		x		x		x			x		6.68 ± 0.21	*		2	DOW 04b	2004
38	39	low	x		x			x				x	2.60 ± 0.89	0.35 ± 0.16	2.2	*	LES 07	2007
39	40	108.0	x		x				x			x	1.61 ± 0.73	0.46 ± 0.26	3.5	4	BOW 06	2006
40	40	110.0	x		x				x			x	2.20 ± 0.69	0.60 ± 0.19	3.7	4	BOW 03	2003

Table 1.2. c) *Wear of metal-on-metal bearings according to [KRE 09]*
(© 2009, with permission from John Wiley & Sons)

#	diameter mm	radial clearance µm	carbon content head hc	carbon content head lc	carbon content cup hc	carbon content cup lc	manufacturing-heat treatment head wr	head ac	head HT	cup wr	cup ac	cup HT	RI wear rate ± SD mm³/10⁶ cyc	SS wear rate ± SD mm³/10⁶ cyc	RI/SS ratio	n	author	year
54	54.5	53.0	x		x			x				x	0.76 ± 0.60	0.09 *	8.4	4	DOW 04a	2004
55	55	low	x		x			x				x	1.20 ± 0.20	0.20 ± 0,04	1.8	*	LES 07	2007
56	56	110.0	x		x				x			x	7.10 ± 2.70	0.40 *	17.8	4	BOW 04	2004
												mean:	2.11	0.36	6.4			

hc	high carbon
lc	low carbon
wr	wrought implants
ac	as cast
HT	heat treated
RI	running-in wear rate
SS	steady-state wear rate
n	sum of tested
(grey square)	head-cup combination with different carbon content or manufacturing method
*	not available

Table 1.2. d) *Wear of metal-on-metal bearings according to [KRE 09] (© 2009, with permission from John Wiley & Sons)*

Table 1.2 contains 56 different investigations including a total of 200 implants. The results are sorted by implant diameter, with radial clearance serving as a secondary sorting criterion. Implants most frequently had a diameter of 28 mm, followed by 36 mm. The 28 mm implants are mainly wrought. The 36 mm implants are evenly split between wrought and cast whereas the larger implants (diameter > 36 mm) are mainly cast. Considering all implants, the radial clearance varies from 22 μm to 160 μm. Table 1.2 clearly shows a decreased steady-state mean wear rate after an initial running-in phase. Considering all the implants from Table 1.2, the mean running-in wear rate is 2.1 $mm^3/10^6$ cycles (SD: ±1.8 $mm^3/10^6$ cycles) and the mean steady-state wear rate is 0.4 $mm^3/10^6$ cycles (SD: ±0.3 $mm^3/10^6$ cycles).

Figure 1.11 outlines simulator studies on implants made from materials other than metal [BOW 08, END 02, OON 04, GAL 08, WIL 03, ESS 05, GAL 05, WAN 03, KUB 09].

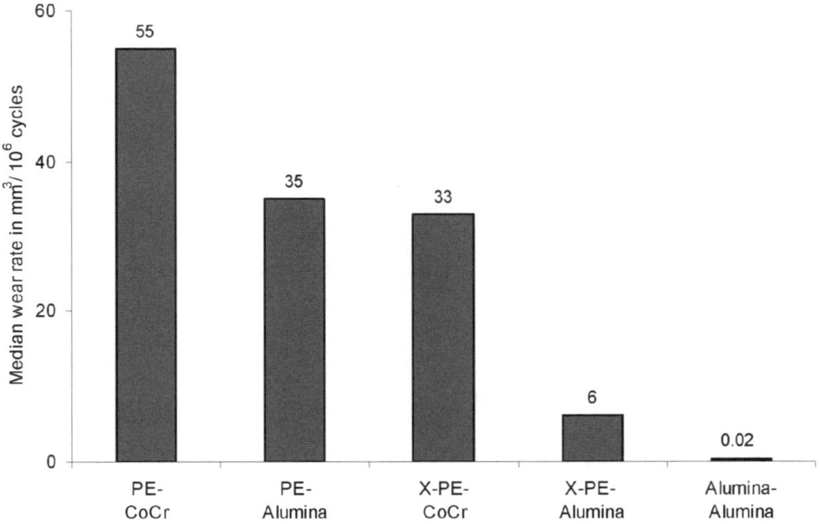

Figure 1.11. *Wear rates of different bearing types. Median wear rate based on multiple simulator studies [BOW 08, END 02, OON 04, GAL 08, WIL 03, ESS 05, GAL 05, WAN 03, KUB 09]*

The wear of metal-on-metal devices is much smaller than the wear of PE bearings and is only outperformed by ceramic-on-ceramic implants (see Figure 1.11).

1.5.5. *Summary of results from simulator studies*

A meta-analysis [KRE 09] of hip simulator wear studies performed during the last decade supports the following statements regarding design- and manufacturing-related parameters and their influence on the wear of metal-on-metal bearings:

– For implants with a diameter of 36 mm and above, an increase in head size leads to reduced running-in wear. If lubrication is insufficient and components are not at least partly separated by a fluid film, wear might increase due to the longer wear path. Thus, sufficient lubrication is essential.

– A smaller clearance reduces running-in wear. However, the minimum clearance is limited and equatorial contact must be avoided.

– A smooth surface (low surface roughness) and a highly spherical geometry (low deviation on roundness) reduce wear.

– The influence of alloy carbon content remains unclear.

– The manufacturing method (wrought vs. cast) does not affect wear.

– Heat treatment processes increase wear, at least during the steady-state wear phase.

A direct comparison of different simulator studies must be read with caution since wear of similar implant designs can differ significantly between different investigators. This is evident when directly comparing the results of Dowson *et al.* [DOW 04a] and Chan *et al.* [CHA 99], who investigated an identical implant design. Dowson *et al.* [DOW 04a] determined a mean running-in wear rate of 2.3 $mm^3/10^6$ cycles, whereas Chan *et al.* reported a running-in wear rate of about one order of magnitude lower (0.24 $mm^3/10^6$ cycles) [CHA 99]. These results show that the comparability of wear studies from different investigators is limited.

Many parameters play a role in minimizing wear and improving outcomes apart from purely technical characteristics of artificial joints. Critical factors include positioning and orientation of implant components during surgery.

Wear of metal-on-metal implants increases with an increasing cup implantation angle [HAR 08]. The risk of impingement (contact of implant neck with implant cup) or luxation (dislocation of the joint) is also influenced by the orientation of the implant components. Additionally, patient-specific aspects such as body weight and activity level are also assumed to impact wear.

1.5.6. *Wear mode*

Different wear modes are used to categorize the wear mechanisms of artificial joints. The four wear modes depend on the bearing partners in use [MCK 07]. This categorization is independent of the bearing partner materials and applies to all artificial joints (including knee, shoulder, etc.). Mode 1 is characterized by articulation of primary bearing surfaces only (see Figure 1.12-1). Although wear takes place, this mode represents the ideal conditions implants are designed for. Modes 2-4 characterize malfunctioning implants. In mode 2, a bearing surface articulates with a secondary non-bearing surface (see Figure 1.12-2). This can be the case when the head of the implant (sub)luxates and subsequently makes contact with the rim of the cup. Mode 2 can lead to massive wear and rapid failure of an artificial joint. Mode 3 describes the articulation of primary bearing surfaces in the presence of third bodies in the joint space, leading to increased abrasive wear (see Figure 1.12-3). Examples of third bodies include metal particles, ceramic fragments, bone cement and bone fragments. Third bodies can increase wear significantly. Mode 4 is characterized by articulation of two non-bearing surfaces (Figure 12-4), as is the case for impingement or "backside" wear. Backside wear can occur in many ways including wear between the acetabular liner and the shell, fretting between modular components, and friction between the implant stem and the surrounding bone or bone cement. Fragments and particles generated in mode 4 can penetrate the joint space and cause mode 3 wear. In combination, mode 2, 3 and 4 wear can lead to massive wear and rapid implant failure, as shown in Figure 1.13 [KLE 09].

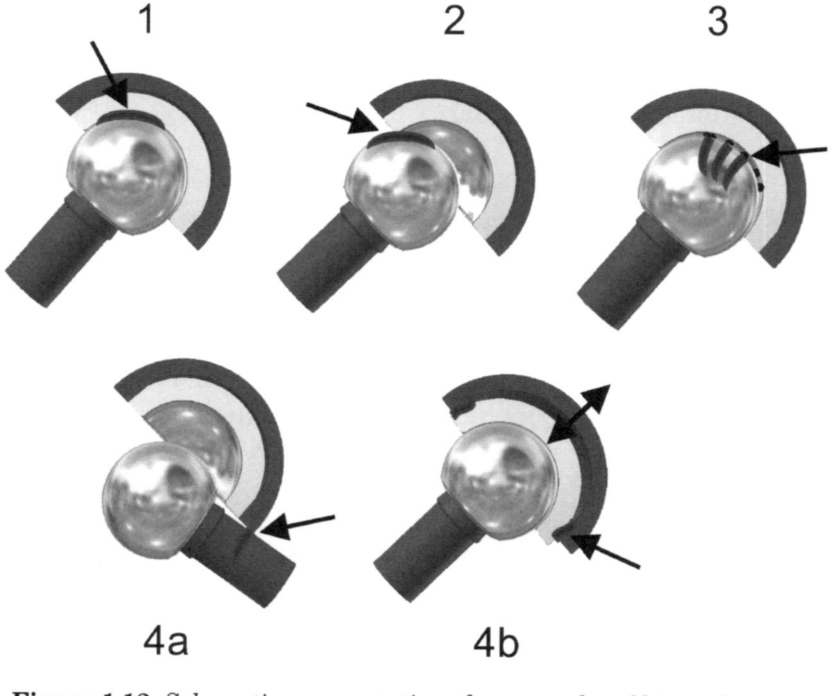

Figure 1.12. *Schematic representation of wear modes of hip replacements based on McKellop [MCK 07] (© 2007, with permission from Elsevier Limited)*

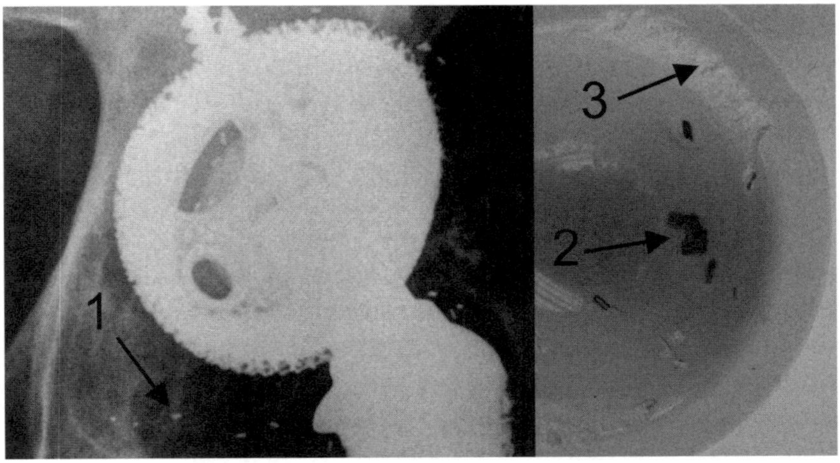

Figure 1.13. *The X-ray (left) shows metallic fragments (1) detached from the macro structure of the cup. These fragments invaded the joint space (2) and caused massive third-body wear (right: retrieved inlay). The edge of the inlay also shows signs of impingement (3)*

1.6. Clinical relevance of metal wear particles and metal ions

Wear of metal-on-metal bearings results in the discharge of metal particles (see Figure 1.14) and metal ions into the tissue and body fluid of patients.

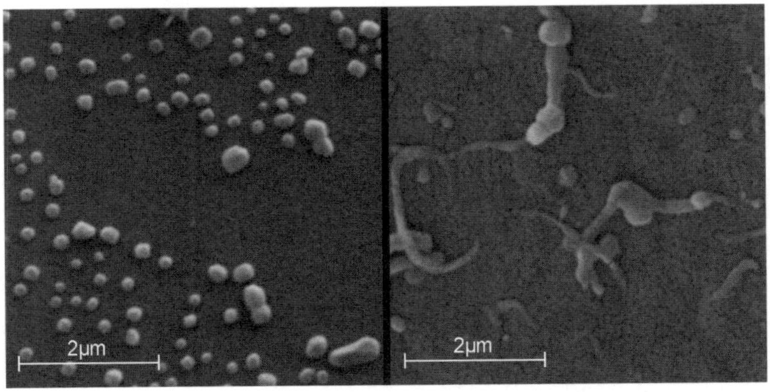

Figure 1.14. *Isolated particles from a simulator study: metal particles (left) and PE particles (right)*

Metal particles are approx. 25-50 nm in size and are mainly spherical or spheroidal in shape. About $6.7x10^{12}$ to $2.5x10^{14}$ particles are released into the surrounding tissue annually [DOO 98]. Thus, a large number of very small uniformly sized and shaped particles are generated from metal-on-metal bearings. Metal particles are susceptible to electrochemical corrosion and hence ion production, sometimes in conjunction with chemically active degradation processes [JAC 03, JAC 98]. In contrast to massive implants, the size of metal particles yields a critical surface to volume ratio and thus a large area for corrosion processes and interactions with body fluids [ING 00]. Consequently, the size, shape and chemical composition of metal particles *in vivo* constantly change [SHA 95].

Chromium and cobalt ions bind to erythrocytes and proteins, especially albumin, the most abundant plasma protein in humans [MER 96, HAL 00]. Histological samples from patients with metal-on-metal implants have revealed a systematic distribution of particles within the body [LAN 92]. Metal particles are not isolated in periprosthetic tissue but rather are transported to distant tissues and organs such as the liver, spleen and bone marrow [CAS 94, URB 00].

Many studies have reported elevated ion concentrations in serum, whole blood and urine of patients with metal-on-metal implants [JAC 98, CLA 03, HEI 05, SKI 02, IAV 06, LHO 03, MAE 04, BRO 03, SAV 03, HEI 08]. Cobalt and molybdenum are cleared by the kidneys relatively fast, whereas chromium is cleared much more slowly and tends to accumulate in the body [MER 96, MER 89]. Possible long-term issues related to particles and ions are cytotoxic or mutagenic reactions [DAL 04, MAL 93, SUN 97, DAV 05], hypersensitivity [HAL 01, GRA 06] and carcinogenicity [VIS 96, MAT 95], all of which are controversially discussed. Single cases of osteolysis have also been reported [KLA 99, BEA 01].

Recently, some reports of inflammatory pseudotumors after metal-on-metal hip surface replacements have been reported [PAN 08, MAH 09, DEP 10, GLY 09]. Pseudotumours are a serious complication defined as a local soft tissue mass associated with localized bone or tissue destruction. The incidence of clinically important pseudotumours is reported to lie in the range of 0.15% [MAL 09] to 6% [DEP 10]. While the risk for men seems to be low, young women are affected much more frequently (up to 13%) [GLY 09]. The reason for the development of pseudotumors is mostly unknown. Current hypotheses cite cytotoxic responses and/or delayed hypersensitivity responses to Co-Cr particles or ions as possible explanations [MAH 09].

1.7. Conclusion

All materials and combinations of materials have potential advantages and disadvantages. Although polymers exhibit higher wear than ceramics or metals, they are more forgiving and less sensitive to poor surgical placement. Due to their hardness, ceramics have lower wear. High loads or point loads (e.g. due to tilted cup inlays) can, however, lead to fracture of the material because no plastic deformation occurs. Metals are wear and fracture resistant. Massive wear can occur in the case of implantation inaccuracies or errors in orientation of components. Metal particles and ions are viewed critically, especially when wear is high.

In spite of the encouraging long-term clinical results for some hip joint replacements, no device exists to date which guarantees lasting durability for young and active patients. Intensive research and development efforts will be required to achieve this goal.

1.8. Acknowledgments

I would like to thank Elsevier Limited, John Wiley & Sons and Professional Engineering Publishing for their kind permission to reuse tables and figures from previously published material [KRE 09, MCK 07, STR 96].

1.9. Bibliography

[AFF 06] AFFATATO S., LEARDINI W., JEDENMALM A., RUGGERI O., TONI A., "Larger diameter bearings reduce wear in metal-on-metal hip implants", *Clinical Orthopaedics and Related Research*, vol. 456, 2006, p. 153-158.

[AMS 96] AMSTUTZ H.C., GRIGORIS P., "Metal on metal bearings in hip arthroplasty", *Clinical Orthopaedics and Related Research*, vol. 329, 1996, p. 11-34.

[AMS 01] AMSTUTZ H.C., "History of metal-on-metal articulations including surface arthroplasty of the hip", in: RIEKER C. (Ed.): *World Tribology Forum in Arthroplasty*, Huber, Bern, 2001, p. 113-123.

[ANG 07] ANGADJI A., ROYLE M., COLLINS S., SHELTON J.C., "Can lubrication theory predict wear in metal-on-metal hip replacements?", *53rd Annual Meeting of the Orthopaedic Research Society*, 2007, San Diego.

[ANI 01] ANISSIAN H.L., STARK A., GOOD V., DAHLSTRAND H., CLARKE I.C., "The wear pattern in metal-on-metal hip prostheses", *Journal of Biomedical Materials Research*, vol. 58, 2001, p. 673-678.

[ANT 07] ANTONACCI J.M., SCHMIDT T.A., SERVENTI L.A., SHU Y.Y.L., GASTELUM N.S., SCHUMACHER B.L., MCILWRAITH C.W., SAH R.L., "Effects of joint injury on synovial fluid and boundary lubrication of cartilage", *53rd Annual Meeting of the Orthopaedic Research Society*, 2007, San Diego.

[BEA 01] BEAULÉ P.E., CAMPBELL P., MIRRA J., HOOPER J.C., SCHMALZRIED T.P., "Osteolysis in a cementless, second generation metal-on-metal hip replacement", *Clinical Orthopaedics and Related Research*, vol. 386, 2001, p. 159-165.

[BER 93] BERGMANN G., GRAICHEN F., ROHLMANN A., "Hip joint loading during walking and running, measured in two patients", *Journal of Biomechanics*, vol. 26, 1993, p. 969-990.

[BER 05] BERRY D.J., "'Minimally invasive' total hip arthroplasty", *Journal of Bone and Joint Surgery* (American), vol. 87, Editorial, 2005, p. 699-700.

[BOW 03] BOWSHER J.G., NEVELOS J., PICKARD J., SHELTON J.C., "Do heat treatments influence the wear of large diameter metal-on-metal hip joints? An in vitro study under normal and adverse gait conditions", *49th Annual Meeting Orthopaedic Research Society*, 2003, New Orleans.

[BOW 04] BOWSHER J.G., HUSSAIN A., NEVELOS J., SHELTON J.C., "The importance of head diameter in minimizing metal-on-metal hip wear", *50th Annual Meeting Orthopaedic Research Society*, 2004, San Francisco.

[BOW 06] BOWSHER J.G., NEVELOS J., WILLIAMS P.A., SHELTON J.C., "'Severe' wear challenge to 'as-cast' and 'double heat-treated' large-diameter metal-on-metal hip bearings", *Proceedings of the Institution of Mechanical Engineers Part H*, vol. 220, 2006, p. 135-143.

[BOW 07] BOWSHER J., DONALDSON J., WANG A., ESSNER A., CLARKE I., "Low running-in wear for large diameter metal-on-metal hip bearings", *53rd Annual Meeting of the Orthopaedic Research Society*, 2007, San Diego.

[BOW 08] BOWSHER J.G., WILLIAMS P.A., CLARKE I.C., GREEN D.D., DONALDSON T.K., "'Severe' wear challenge to 36 mm mechanically enhanced highly crosslinked polyethylene hip liners", *Journal of Biomedical Materials Research Part B: Applied Biomaterials*, vol. 86 no. 1, July 2008, p. 253-63.

[BRI 98] BRISCOE B.J., "Isolated contact stress deformations of polymers: the basis for interpreting polymer tribology", *Tribology International*, vol. 31, 1998, p. 121-126.

[BRO 03] BRODNER W., BITZAN P., MEISINGER V., KAIDER A., GOTTSAUNER-WOLF F., KOTZ R., "Serum cobalt levels after metal-on-metal total hip arthroplasty", *Journal of Bone and Joint Surgery* (American), vol. 85, 2003, p. 2168-2173.

[BRO 07] BROCKETT C.L., HARPER P., WILLIAMS S., ISAAC G., DWYER-JOYCE R., JIN Z., FISHER J., "Direct measurement of lubrication in large diameter metal-on-metal hip implants", *53rd Annual Meeting of the Orthopaedic Research Society*, 2007, San Diego.

[CAL 07] CALIGARIS M., ATESHIAN G., "Cartilage interstitial fluid pressurization is far more effective at reducing friction than boundary lubrication by synovial fluid", *53rd Annual Meeting of the Orthopaedic Research Society*, 2007, San Diego.

[CAS 94] CASE C.P., LANGKAMER V.G., JAMES C., PALMER M.R., KEMP A.J., HEAP P.F., SOLOMON L., "Widespread dissemination of metal debris from implants", *Journal of Bone and Joint Surgery* (British), vol. 76, 1994, p. 701-712.

[CAU 02] CAUDILLO M., HERRERA-TREJO M., CASTRO M.R., RAMIREZ E., GONZALEZ C.R., JUAREZ J.I., "On carbide dissolution in an as-cast ASTM F-75 alloy", *Journal of Biomedical Materials Research*, vol. 59, 2002, p. 378-385.

[CAW 03] CAWLEY J., METCALF J.E., JONES A.H., BAND T.J., SKUPIEN D.S., "A tribological study of cobalt chromium molybdenum alloys used in metal-on-metal resurfacing hip Arthroplasty", *Wear*, vol. 255, 2003, p. 999-1006.

[CHA 60] CHARNLEY J., "The lubrication of animal joints in relation to surgical reconstruction by arthroplasty", *Annals of the Rheumatic Diseases*, vol. 19, 1960, p. 10-19.

[CHA 61] CHARNLEY J., "Arthroplasty of the hip. A new operation", *The Lancet*, vol. 1, 1961, p. 1129-1132.

[CHA 96] CHAN F.W., BOBYN J.D., MEDLEY J.B., KRYGIER J.J., YUE S., TANZER M., "Engineering issues and wear performance of metal on metal hip implants", *Clinical Orthopaedics and Related Research*, vol. 333, 1996, p. 96-107.

[CHA 99] CHAN F.W., BOBYN J.D., MEDLEY J.B., KRYGIER J.J., TANZER M., "The Otto Aufranc Award. Wear and lubrication of metal-on-metal hip implants", *Clinical Orthopaedics and Related Research*, vol. 369, 1999, p. 10-24.

[CLA 99] CLARK J.M., NORMAN A.G., KAAB M.J., NOTZLI H.P., "The surface contour of articular cartilage in an intact, loaded joint", *Journal of Anatomy*, vol. 195, 1999, p. 45-56.

[CLA 03] CLARKE M.T., LEE P.T., ARORA A., VILLAR R.N., "Levels of metal ions after small- and large-diameter metal-on-metal hip arthroplasty", *Journal of Bone and Joint Surgery* (British), vol. 85, 2003, p. 913-917.

[CZI 03] CZICHOS H., HABIG K.H., *Tribologie Handbuch, Reibung und Verschleiß*, 2nd Edition, Vieweg Wiesbaden, 2003, p. 225-226.

[DAL 04] DALEY B., DOHERTY A.T., FAIRMAN B., CASE C.P., "Wear debris from hip or knee replacements causes chromosomal damage in human cells in tissue culture", *Journal of Bone and Joint Surgery* (British), vol. 86, 2004, p. 598-606.

[DAV 05] DAVIES A.P., SOOD A., LEWIS A.C., NEWSON R., LEARMONTH I.D., CASE C.P., "Metal-specific differences in levels of DNA damage caused by synovial fluid recovered at revision arthroplasty", *Journal of Bone and Joint Surgery* (British), vol. 87, 2005, p. 1439-1444.

[DEP 10] DEPREZ P., SINT-KRUIS L.V.B., DEMUYNCK M., "High early revision rate due to pseudotumor formation in Metal-on-Metal Large Head Diameter THA", *AAOS Annual Meeting*, 09-13 March 2010, New Orleans.

[DOB 83] DOBBS H.S., ROBERTSON J.L., "Heat treatment of cast Co-Cr-Mo for orthopaedic implant use", *Journal of Materials Science*, vol. 18, 1983, p. 391-401.

[DOO 98] DOORN P.F., CAMPBELL P.A., WORRALL J., BENYA P.D., MCKELLOP H.A., AMSTUTZ H.C., "Metal wear particle characterization from metal on metal total hip replacements: transmission electron microscopy study of periprosthetic tissues and isolated particles", *Journal of Biomedical Materials Research*, vol. 42, 1998, p. 103-111.

[DOW 00] DOWSON D., MCNIE C.M., GOLDSMITH A.A.J., "Direct experimental evidence of lubrication in a metal-on-metal total hip replacement tested in a joint simulator", *Proceedings of the Institution of Mechanical Engineers Part H*, vol. 214, 2000, p. 75-86.

[DOW 04a] DOWSON D., HARDAKER C., FLETT M., ISAAC G.H. "A hip joint simulator study of the performance of metal-on-metal joints: Part II: design", *Journal of Arthroplasty*, vol. 19, 2004, p. 124-130.

[DOW 04b] DOWSON D., HARDAKER C., FLETT M., ISAAC G.H., "A hip joint simulator study of the performance of metal-on-metal joints: Part I: the role of materials", *Journal of Arthroplasty*, vol. 19, 2004, p. 118-123.

[DOW 06] DOWSON D., JIN Z.M., "Metal-on-metal hip joint tribology", *Proceedings of the Institution of Mechanical Engineers Part H*, vol. 220, 2006, p. 107-118.

[END 02] ENDO M., TIPPER J.L., BARTON D.C., STONE M.H., INGHAM E., FISHER J., "Comparison of wear, wear debris and functional biological activity of moderately crosslinked and non-crosslinked polyethylenes in hip prostheses", *Proceedings of the Institution of Mechanical Engineers Part H*, vol. 216 no. 2, 2002, p. 111-22.

[ESS 05] ESSNER A., SUTTON K., WANG A., "Hip simulator wear comparison of metal-on-metal, ceramic-on-ceramic and crosslinked UHMWPE bearings", *Wear*, vol. 259, 2005, p. 992-995.

[FIR 01] FIRKINS P.J., TIPPER J.L., SAADATZADEH M.R., INGHAM E., STONE M.H., FARRAR R., FISHER J., "Quantitative analysis of wear and wear debris from metal-on-metal hip prostheses tested in a physiological hip joint simulator", *Biomedical Materials and Engineering*, vol. 11, 2001, p. 143-157.

[FIS 02] FISHER J., HU X.Q., TIPPER J.L., STEWART T.D., WILLIAMS S., STONE M.H., DAVIES C., HATTO P., BOLTON J., RILEY M., HARDAKER C., ISAAC G.H., BERRY G., INGHAM E., "An in vitro study of the reduction in wear of metal-on-metal hip prostheses using surface-engineered femoral heads", *Proceedings of the Institution of Mechanical Engineers Part H*, vol. 216, 2002, p. 219-230.

[FIS 04] FISHER J., HU X.Q., STEWART T.D., WILLIAMS S., TIPPER J.L., INGHAM E., STONE M.H., DAVIES C., HATTO P., BOLTON J., RILEY M., HARDAKER C., ISAAC G.H., BERRY G, "Wear of surface engineered metal-on-metal hip prostheses", *Journal of Materials Science: Materials in Medicine*, vol. 15, 2004, p. 225-235.

[GAL 05] GALVIN A.L., WILLIAMS S., HATTO P., THOMPSON J., ISAAC G., STONE M., INGHAM E., FISHER J., "Comparison of wear of ultra high molecular weight polyethylene acetabular cups against alumina ceramic and chromium nitride coated femoral heads", *Wear*, vol. 259, 2005, p. 972-976.

[GAL 08] GALVIN A., BROCKETT C., WILLIAMS S., HATTO P., BURTON A., ISAAC G., STONE M., INGHAM E., FISHER J., "Comparison of wear of ultra-high molecular weight polyethylene acetabular cups against surface-engineered femoral heads", *Proceedings of the Institution of Mechanical Engineers Part H*, vol. 222 no. 7, October 2008, p. 1073-80.

[GLY 09] GLYN-JONES S., PANDIT H., KWON Y.M., DOLL H., GILL H.S., MURRAY D.W., "Risk factors for inflammatory pseudotumour formation following hip resurfacing", *Journal of Bone and Joint Surgery* (British), vol. 91, no. 12, December 2009, p. 1566-1574.

[GOL 00] GOLDSMITH A.A., DOWSON D., ISAAC G.H., LANCASTER J.G., "A comparative joint simulator study of the wear of metal-on-metal and alternative material combinations in hip replacements", *Proceedings of the Institution of Mechanical Engineers Part H*, vol. 214, 2000, p. 39-47.

[GRA 06] GRANCHI D., CENNI E., TRISOLINO G., GIUNTI A., BALDINI N., "Sensitivity to implant materials in patients undergoing total hip replacement", *Journal of Biomedical Materials Research Part B: Applied Biomaterials*, vol. 77, 2006, p. 257-264.

[HAL 00] HALLAB N.J., JACOBS J.J., SKIPOR A., BLACK J., MIKECZ K., GALANTE J.O., "Systemic metal-protein binding associated with total joint replacement arthroplasty", *Journal of Biomedical Materials Research*, vol. 49, 2000, p. 353-361.

[HAL 01] HALLAB N.J., MERRITT K., JACOBS J.J., "Metal sensitivity in patients with orthopaedic implants", *Journal of Bone and Joint Surgery* (American), vol. 83, 2001, p. 428-436.

[HAM 78] HAMROCK B.J., DOWSON D., "Elsastohydrodynamic lubrication of elliptical contacts for materials of low elastic modulus I – Fully flooded conjunction", *Journal of Lubrication Technology*, vol. 100, 1978, p. 236-245.

[HAR 06] HARDAKER C., DOWSON D., ISAAC G.H., "Head replacement, head rotation, and surface damage effects on metal-on-metal total hip replacements: a hip simulator study", *Proceedings of the Institution of Mechanical Engineers Part H*, vol. 220, 2006, p. 209-217.

[HAR 08] HART A.J., BUDDHDEV P., WINSHIP P., FARIA N., POWELL J.J., SKINNER J.A., "Cup inclination angle of greater than 50 degrees increases whole blood concentrations of cobalt and chromium ions after metal-on-metal hip resurfacing", *Hip International*, vol. 18 no. 3, July-September 2008, 212-219.

[HEI 05] HEISEL C., SILVA M., SKIPOR A.K., JACOBS J.J., SCHMALZRIED T.P., "The relationship between activity and ions in patients with metal-on-metal bearing hip prostheses", *Journal of Bone and Joint Surgery* (American), vol. 87, 2005, p. 781-787.

[HEI 08] HEISEL C., STREICH N., KRACHLER M., JAKUBOWITZ E., KRETZER J.P., "Characterization of the running-in period in total hip resurfacing arthroplasty: an in vivo and in vitro metal ion analysis", *Journal of Bone and Joint Surgery* (American), vol. 90 suppl. 3, August 2008, p. 125-33.

[HEI 09] HEISEL C., KLEINHANS J.A., MENGE M., KRETZER J.P., "Ten different hip resurfacing systems: biomechanical analysis of design and material properties", *International Orthopaedics*, vol. 33 no. 4, August 2009, p. 939-943. Epub 4 July 2008.

[IAV 06] IAVICOLI I., FALCONE G., ALESSANDRELLI M., CRESTI R., DE SANTIS V., SALVATORI S., ALIMONTI A., CARELLI G., "The release of metals from metal-on-metal surface arthroplasty of the hip", *Journal of Trace Elements in Medicine and Biology*, vol. 20, 2006, p. 25-31.

[ING 00] INGHAM E., FISHER J., "Biological reactions to wear debris in total joint replacement", *Proceedings of the Institution of Mechanical Engineers Part H*, vol. 214, 2000, p. 21-37.

[JAC 98] JACOBS J.J., SKIPOR A.K., PATTERSON L.M., HALLAB N.J., PAPROSKY W.G., BLACK J., GALANTE J.O., "Metal release in patients who have had a primary total hip arthroplasty. A prospective, controlled, longitudinal study", *Journal of Bone and Joint Surgery* (American), vol. 80, 1998, p. 1447-1458.

[JAC 03] JACOBS J.J., HALLAB N.J., SKIPOR A.K., URBAN R.M., "Metal degradation products: a cause for concern in metal-metal bearings?", *Clinical Orthopaedics and Related Research*, vol. 417, 2003, p. 139-147.

[JAL 01] JALALI-VAHID D., JAGATIA M., JIN Z.M., DOWSON D., "Prediction of lubricating film thickness in UHMWPE hip joint replacements", *Journal of Biomechanics*, vol. 34, 2001, p. 261-266.

[JAL 06] JALALI-VAHID D., JIN Z.M., DOWSON D., "Effect of start-up conditions on elastohydrodynamic lubrication of metal-on-metal hip implants", *Proceedings of the Institution of Mechanical Engineers Part H*, vol. 220, 2006, p. 143-150.

[JIN 93] JIN Z.M., DOWSON D., FISHER J., "Fluid film lubrication in natural hip joints", 1993, p. 545-555, in: DOWSON D. *et al.* (Hrsg): *Thin Films in Tribology*. Elsevier, Amsterdam.

[JIN 97] JIN Z.M., DOWSON D., FISHER J., "Analysis of fluid film lubrication in artificial hip joint replacements with surfaces of high elastic modulus", *Proceedings of the Institution of Mechanical Engineers Part H*, vol. 211, 1997, p. 247-256.

[JOH 99] JOHN K.R., POGGIE R.A., ZARDIACKAS L.D., AFFLITTO R.M., "Comparison of two cobalt-based alloys for use in metal-on-metal hip prostheses: Evaluation of the wear properties in a simulator", 1999, p. 145-155, in: DISEGI J.A., KENNEDY R.L., PILLIAR R. (Ed.): *Cobalt-base Alloys for Biomedical Applications*, ASTM, West Conshohocken.

[KÄR 08] KÄRRHOLM J., GARELLICK G., ROGMARK C., HERBERTS P., Swedish Hip Arthroplasty Register, Annual Report 2007. Gothenburg, Sweden, 2008. http://www.jru.orthop.gu.se/archive /AnnualReport-2007-eng.pdf (date last accessed 14 April 2010).

[KLA 99] KLAPPERICH C., GRAHAM J., PRUITT L., RIES M.D., "Failure of a metal-on-metal total hip arthroplasty from progressive osteolysis", *Journal of Arthroplasty*, vol. 14, 1999, p. 877-881.

[KLE 09] KLEINHANS J.A., JAKUBOWITZ E., SEEGER J.B., HEISEL C., KRETZER J.P., "Macroscopic third-body wear caused by porous metal surface fragments in total hip arthroplasty", *Orthopedics*, vol. 32 no. 5, May 2009, p. 364.

[KRE 09] KRETZER J.P., KLEINHANS J.A., JAKUBOWITZ E., THOMSEN M., HEISEL C., "A meta-analysis of design- and manufacturing-related parameters influencing the wear behavior of metal-on-metal hip joint replacements", *Journal of Orthopaedic Research*, vol. 27 no. 11, November 2009, p. 1473-1480.

[KRE 10] KRETZER J.P., KRACHLER M., REIDERS J., JAKUBOWITZ E., THOMSEN M., HEISEL C., "Determination of low wear rates in metal-on-metal hip joint replacements based on ultra trace element analysis in simulator studies", *Tribology Letters*, no. 37, January 2010, p. 23-29.

[KUB 09] KUBO K., CLARKE I.C., SORIMACHI T., WILLIAMS P.A., DONALDSON T.K., YAMAMOTO K., "Aggressive 3rd-body wear challenge to highly crosslinked polyethylene: A hip simulator model", *Wear*. vol. 267, 2009, p. 734-42.

[LAN 92] LANGKAMER V.G., CASE C.P., HEAP P., TAYLOR A., COLLINS C., PEARSE M., SOLOMON L., "Systemic distribution of wear debris after hip replacement. A cause for concern?", *Journal of Bone and Joint Surgery* (British), vol. 74, 1992, p. 831-839.

[LEE 06] LEE S.H., TAKAHASHI E., NOMURA N., CHIBA A., "Effect of carbon addition on microstructure and mechanical properties of a wrought Co-Cr-Mo implant alloy", *Materials Transactions*, vol. 47, 2006, p. 287-290.

[LES 07] LESLIE I.J., WILLIAMS S., BROWN C., THOMPSON J., ISAAC G., INGHAM E., FISHER J., "Effect of bearing size on wear, wear debris and ion levels of surface replacements", *53rd Annual Meeting of the Orthopaedic Research Society*, 2007, San Diego.

[LHO 03] LHOTKA C., SZEKERES T., STEFFAN I., ZHUBER K., ZWEYMULLER K., "Four-year study of cobalt and chromium blood levels in patients managed with two different metal-on-metal total hip replacements", *Journal of Orthopaedic Research*, vol. 21, 2003, p. 189-195.

[LIA 04] LIAO Y.S., FRYMANN C., HANES M., TURNER L., "Effects of clearance, head size and start-stop protocol on wear of metal-on-metal hip joint bearings in physiological anatomical hip joint simulator", *50th Annual Meeting Orthopaedic Research Society*, 2004, San Francisco.

[LIA 07] LIAO Y.S., "Wear reduction of large head metal-on-metal implants in a hip simulation study", *53rd Annual Meeting of the Orthopaedic Research Society*, 2007, San Diego.

[MAE 04] MAEZAWA K., NOZAWA M., MATSUDA K., YASUMA M., MORI K., ENOMOTO F., OGAWA S., SHITOTO K., KUROSAWA H., "Chronological changes of serum chromium levels after modern metal-on-metal total hip arthroplasty", *Acta Orthop Scand*, vol. 75, 2004, p. 422-526.

[MAH 09] MAHENDRA G., PANDIT H., KLISKEY K., MURRAY D., GILL H.S., ATHANASOU N., "Necrotic and inflammatory changes in metal-on-metal resurfacing hip arthroplasties", *Acta Orthopaedica*, vol. 80 no. 6, December 2009, p. 653-659.

[MAL 93] MALONEY W.J., SMITH R.L., CASTRO F., SCHURMAN D.J., "Fibroblast response to metallic debris in vitro. Enzyme induction cell proliferation, and toxicity", *Journal of Bone and Joint Surgery* (American), vol. 75, 1993, p. 835-844.

[MAL 09] MALVIYA A., HOLLAND J.P., "Pseudotumours associated with metal-on-metal hip resurfacing: 10-year Newcastle experience", *Acta Orthopaedica Belgica*, vol. 75 no. 4, August 2009 p. 477-483.

[MAT 95] MATHIESEN E.B., AHLBOM A., BERMANN G., LINDGREN J.U., "Total hip replacement and cancer. A cohort study", *Journal of Bone and Joint Surgery* (British), vol. 77, 1995, p. 345-350.

[MCK 07] MCKELLOP H.A., "The lexicon of polyethylene wear in artificial joints", *Biomaterials*, vol. 28 no. 34, December 2007, p. 5049-5057. Epub 16 August 2007.

[MED 98] MEDLEY J.B., DOWLING J.M., POGGIE R.A., KRYGIER J.J., BOBYN J.D., "Simulator wear of some commercially available metal-on-metal hip implants", 1998, p. 92-110, in: JACOBS J.J., CRAIG T.L. (Ed.): *Alternative Bearing Surfaces in Total Joint Replacement*, ASTM, West Conshohocken.

[MED 01] MEDLEY J.B., BOBYN J.D., KRYGIER J.J., CHAN F.W., TANZER M., ROTER G.E., "Elastohydrodynamic lubrication and wear of metal-on-metal hip implants", 2001, p. 125-136, in: RIEKER C. (Ed.): *World Tribology Forum in Arthroplasty*. Huber, Bern.

[MER 89] MERRITT K., CROWE T.D., BROWN S.A., "Elimination of nickel, cobalt, and chromium following repeated injections of high dose metal salts", *Journal of Biomedical Materials Research*, vol. 23, 1989, p. 845-862.

[MER 96] MERRITT K., BROWN S.A., "Distribution of cobalt chromium wear and corrosion products and biologic reactions", *Clinical Orthopaedics and Related Research*, vol. 329, 1996, p. 233-243.

[MET 04] METCALF J.E.P., CAWLEY J., BAND T.J., "Cobalt Chromium Molybdenum metal-on-metal resurfacing orthopaedic hip devices", *Business Briefing: Medical Device Manufacturing & Technology 2004*, 2004. URL: http://www.touchbriefings.com/pdf/954/ACF3EFE.pdf, (last accessed 03 April 2007, 4:34 p.m.)

[MOA 03] MOA-ANDERSON B.J., COSTA K.D., HUNG C.T., ATESHIAN G.A., "Bovine articular cartilage surface topography and roughness in fresh versus frozen tissue samples using atomic force microscopy", *2003 Summer Bioengineering Conference*; 2003, Key Biscanye, Florida.

[NEV 04] NEVELOS J., SHELTON J.C., FISHER J., "Metallurgical considerations in the wear of metal-on-metal hip bearings", *Hip International*, vol. 14, 2004, p. 1-10.

[OON 04] OONISHI H., CLARKE I.C., GOOD V., AMINO H., UENO M., "Alumina hip joints characterized by run-in wear and steady-state wear to 14 million cycles in hip-simulator model", *Journal of Biomedical Materials Research Part A*, vol. 70 no. 4, 15 September 2004, p. 523-32.

[PAN 08] PANDIT H., GLYN-JONES S., MCLARDY-SMITH P., GUNDLE R., WHITWELL D., GIBBONS C.L., OSTLERE S., ATHANASOU N., GILL H.S., MURRAY D.W., "Pseudotumours associated with metal-on-metal hip resurfacings", *Journal of Bone and Joint Surgery* (British), vol. 90 no. 7, July 2008, p. 847-851.

[SAI 97] SAIKKO V., AHLROOS T., "Phospholipids as boundary lubricants in wear tests of prosthetic joint materials", *Wear,* vol. 207, 1997, p. 86-91.

[SAI 05] SAIKKO V., "A 12-station anatomic hip joint simulator", *Proceedings of the Institution of Mechanical Engineers Part H*, vol. 219, 2005, p. 437-448.

[SAV 03] SAVARINO L., GRANCHI D., CIAPETTI G., CENNI E., GRECO M., ROTINI R., VERONESI C.A., BALDINI N., GIUNTI A., "Ion release in stable hip arthroplasties using metal-on-metal articulating surfaces: a comparison between short- and medium-term results", *Journal of Biomedical Materials Research*, vol. 66, 2003, p. 450-456.

[SCH 96] SCHMIDT M., WEBER H., SCHON R., "Cobalt chromium molybdenum metal combination for modular hip prostheses", *Clinical Orthopaedics and Related Research*, vol. 329, 1996, p. 35-47.

[SCH 00] SCHOLES S.C., UNSWORTH A., "Comparison of friction and lubrication of different hip prostheses", *Proceedings of the Institution of Mechanical Engineers Part H*, vol. 214, 2000, p. 49-57.

[SCH 01] SCHOLES S.C., GREEN S.M., UNSWORTH A., "The wear of metal-on-metal total hip prostheses measured in a hip simulator", *Proceedings of the Institution of Mechanical Engineers Part H*, vol. 215, 2001, p. 523-530.

[SHA 95] SHAHGALDI B.F., HEATLEY F.W., DEWAR A., CORRIN B., "In vivo corrosion of cobalt-chromium and titanium wear particles", *Journal of Bone and Joint Surgery* (British), vol. 77, 1995, p. 962-966.

[SHI 93] SHI L., CAO Z.M., "Alloy design and microstructure of a biomedical Co-Cr alloy", *Journal of Materials Science*, vol. 28, 1993, 1312-1316.

[SKI 02] SKIPOR A.K., CAMPBELL P.A., PATTERSON L.M., ANSTUTZ H.C., SCHMALZRIED T.P., JACOBS J.J., "Serum and urine metal levels in patients with metal-on-metal surface arthroplasty", *Journal of Materials Science: Materials in Medicine*, vol. 13, 2002, p. 1227-1234.

[SMI 01] SMITH S.L., DOWSON D., GOLDSMITH A.A., "The effect of femoral head diameter upon lubrication and wear of metal-on-metal total hip replacements", *Proceedings of the Institution of Mechanical Engineers Part H*, vol. 215, 2001, p. 161-170.

[STR 96] STREICHER R.M., SEMLITSCH M., SCHON R., WEBER H., RIEKER C., "Metal-on-metal articulation for artificial hip joints: laboratory study and clinical results", *Proceedings of the Institution of Mechanical Engineers Part H*, vol. 210, 1996, p. 223-232.

[SUN 97] SUN Z.L., WATAHA J.C., HANKS C.T., "Effects of metal ions on osteoblast-like cell metabolism and differentiation", *Journal of Biomedical Materials Research*, vol. 34, 1997, p. 29-37.

[TIP 05] TIPPER J.L., INGHAM E., JIN Z.M., FISHER J., "The science of metal-on-metal articulation", *Current Orthopaedics*, vol. 19, 2005, p. 280-287.

[UNS 91] UNSWORTH A., "Tribology of human and artificial joints", *Proceedings of the Institution of Mechanical Engineers Part H*, vol. 205, 1991, p. 163-172.

[URB 00] URBAN R.M., JACOBS J.J., TOMLINSON M.J., GAVRILOVIC J., BLACK J., PEOC'H M., "Dissemination of wear particles to the liver, spleen, and abdominal lymph nodes of patients with hip or knee replacement", *Journal of Bone and Joint Surgery* (American), vol. 82, 2000, p. 457-476.

[VAR 03] VARANO R., BODYN J.D., MEDLEY J.B., YUE S., "Does alloy heat treatment influence metal-on-metal wear?", *49th Annual Meeting Orthopaedic Research Society*, 2003, New Orleans.

[VAS 06] VASSILIOU K., ELFICK A.P., SCHOLES S.C., UNSWORTH A., "The effect of 'running-in' on the tribology and surface morphology of metal-on-metal Birmingham hip resurfacing device in simulator studies", *Proceedings of the Institution of Mechanical Engineers Part H*, vol. 220, 2006, p. 269-277.

[VIS 87] VISURI T., "Long-term results and survivorship of the McKee-Farrar total hip prosthesis", *Archives of Orthopaedic and Trauma Surgery*, vol. 106, 1987, p. 368-374.

[VIS 96] VISURI T., PUKKALA E., PAAVOLAINEN P., PULKKINEN P., RISKA E.B., "Cancer risk after metal on metal and polyethylene on metal total hip arthroplasty", *Clinical Orthopaedics and Related Research*, vol. 329, 1996, p. 280-289.

[WAN 03] WANG A., SCHMIDIG G., "Ceramic femoral heads prevent runaway wear for highly crosslinked polyethylene acetabular cups by third-body bone cement particles", *Wear*, vol. 255, 2003, p. 1057-1063.

[WEB 96] WEBER B.G., "Experience with the Metasul total hip bearing system", *Clinical Orthopaedics and Related Research*, vol. 329, 1996, p. 69-77.

[WID 02] WIDMER M.R., Modified molecular friction in artificial hip joints, technical dissertation, 2002, ETH Zürich.

[WIL 00] WILLERT H.G., BUCHHORN G.H., FAYYAZI A., LOHMANN C.H., "Histopathologische Veränderungen bei Metall/Metall-Gelenken geben Hinweise auf eine zellvermittelte Überempfindlichkeit", *Osteologie*, vol. 9, 2000, p. 165-179.

[WIL 03] WILLIAMS S., BUTTERFIELD M., STEWART T., INGHAM E., STONE M., FISHER J., "Wear and deformation of ceramic-on-polyethylene total hip replacements with joint laxity and swing phase microseparation", *Proceedings of the Institution of Mechanical Engineers Part H*, vol. 217 no. 2, 2003, p. 147-53.

[WIL 96] WILLMANN G., "Hüftgelenkersatz – eine tribologische und konstruktive Herausforderung", *Materialwissenschaft und Werkstofftechnik*, vol. 27, 1996, p. 199-205.

[ZHA 98] ZHANG S.W., "State-of-the-art of polymer tribology", *Tribology International*, vol. 31, 1998, p. 49-60.

Chapter 2

Experimental Wear Studies of Total Joint Replacements

This chapter outlines the experimental methods used to evaluate the tribology of total joint replacement. It details wear and friction simulation, focusing primarily on studies of the knee and hip.

2.1. Introduction

Total joint replacement must be considered one of the most successful surgical interventions of current medical practice, with the clinical performance of total hip replacement being particularly notable. Historically, patients requiring joint replacement have tended to be older, with a limited range of motion, reduced activity and a shorter life expectancy. These patients have generally been satisfied to have reduced activity after replacement surgery, whilst benefiting from the reduction or elimination of pain. As the patient demographic has changed, with younger and more active patients requiring joint replacement, patients who will

---

Chapter written by Claire BROCKETT and John FISHER.

live longer and participate in more demanding activities, the bearings which have historically been successful may no longer be appropriate. New bearings may be required to enhance the tribological lifetime of a replacement joint, through improved lubrication, reduced friction and reduced wear, whilst providing good functionality. Most importantly, it has been found necessary to reduce the wear rate and rate of generation of wear debris that can accumulate in periprosthetic tissues over time, and reach critical levels which result in adverse tissue reactions, loosening and failure. This is difficult to achieve and requires consideration of design, manufacturing technique and material to improve the performance of the bearing.

In the laboratory it is possible to represent conditions which may occur clinically, to examine the tribological performance of total joint replacement. The laboratory preclinical simulations also provide an essential technology platform in which to assess the safety, efficacy and performance of joint replacements, so reducing need for animal and clinical studies. The next few sections detail the methods used to assess the implants, and the factors that influence the tribology of total joint replacement.

2.2. Methods for assessing tribology in total joint replacement

There have been many studies examining the tribology of total joint replacement. Many lubrication and friction studies focus only on the total hip replacement, or hip resurfacing, but wear studies have been conducted on many total joint replacement designs. Most of these studies have utilized experimental simulators to study the wear and friction of the total joint replacement. It should be noted that when reviewing the literature, there are a broad range of simulators, with different test sample arrangements, and this may influence the data; therefore direct comparison between tests is not always possible.

2.2.1. *Lubrication*

Few techniques have been employed to directly measure the lubrication of total joint replacement, with most studies employing friction assessment to indirectly assess lubrication through theoretical Stribeck analysis. Two methods for measurement of lubrication have been employed on metal-on-metal total hip and resurfacing replacements, with limited success. An electrical resistance technique has been used to measure the lubrication of total hip replacements *in vitro* [GOL 00, MCN 00]. The bearing was arranged such that both the acetabular cup and femoral components were electrically isolated. A circuit set up across the bearing allowed measurement of potential drop across the components. Complete separation of the head and cup by a layer of lubricant was indicated by no passing of current, when current was measured it indicated contact between the head and cup. However, it should be noted that this method only allows qualitative analysis of lubrication, as it gives no indication of separation distance (film thickness) or the level of contact between the head and cup.

The second technique involves ultrasonic measurement of lubrication [BRO 08]. A single piezo-electric sensor was bonded onto the back of a cup, over the area of anticipated minimum film thickness. Reference signals were recorded, without lubricant in the cup, and these were compared to the signals measured during the study. This method relied upon reflection of ultrasonic pulses emitted by the sensor at the lubricant-bearing interface, and was limited to materials that had significantly different acoustic properties to the lubricant. At this stage, it is not possible to assess lubrication of polymeric materials with this technique.

2.2.2. *Friction*

Friction studies on total hip replacements, and natural joints, have been performed for many years. These studies are useful in two ways; measurement of frictional torque gives a good indication for whether the bearing will have stable fixation under standard conditions, as failures of metal-on-metal hip replacements historically have been related to high frictional torques generated due to poor tolerances and clearance. Additionally, measurement of the friction factor, or coefficient, may be used to give a good indication of the lubrication of the bearing.

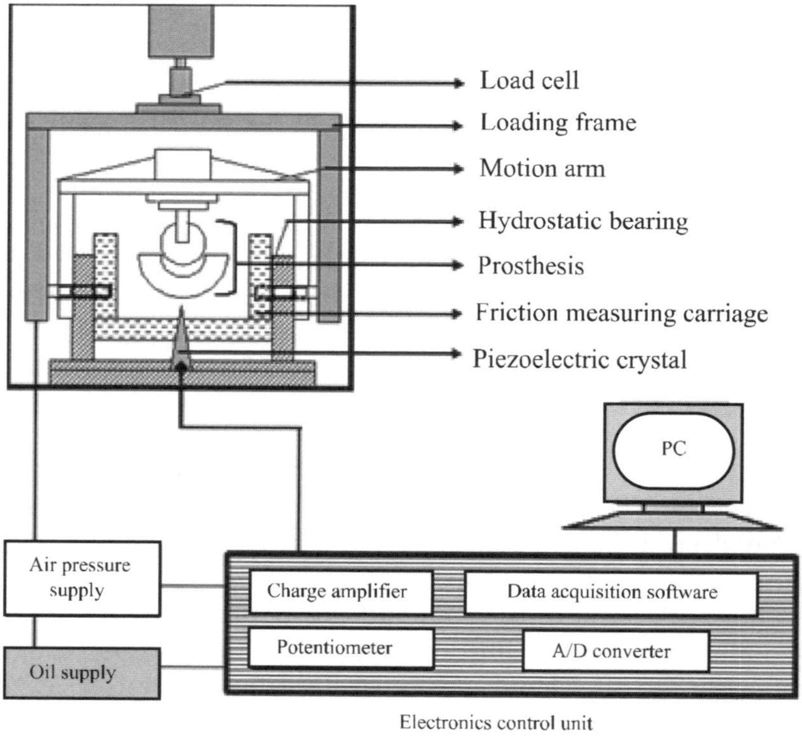

Figure 2.1. *Schematic diagram of a driven pendulum friction simulator [UDO 03]*

Early friction studies employed a simple pendulum machine [CHA 66], where the friction of a joint was determined by assembling the implant as the fulcrum of the pendulum, attaching a defined mass to the pendulum and allowing the bearing to swing freely from a fixed starting position whilst monitoring the decay of the swing. Although this was useful as a qualitative method for comparing different bearings, no quantitative value for friction was derived. This method is based on the theory that the friction occurring between the head and cup will act as a damping force, causing decay of the swing and ultimately stopping the motion of the pendulum.

More recently, driven pendulum simulators have been used to investigate a variety of different materials, designs and factors which may affect the friction of total joint replacements, in both the hip and knee [SCH 00a, SCH 00b, BRO 07]. These simulators are assembled such that a defined load and motion cycle may be applied to the implant for a chosen number of cycles. The loading frame, which holds the test implant, is mounted on a hydrostatic bearing, essentially eliminating the friction of the machine from the measurement. Thus all friction measured by the piezo-electric transducer, attached to the frame (Figure 2.1) can be assumed to be between the two bearing surfaces. In many studies, the bearings have been inverted with respect to anatomical position, but due to the short-term nature of these studies, this is not believed to have a significant effect. The frictional torque (T) measured during these studies can be used to determine the friction factor (f; equation [2.1]);

$$f = \frac{T}{RW} \qquad [2.1]$$

The friction factor defined is similar in magnitude to the coefficient of friction, but is variable with finite contact area; R is the bearing radius and W is the applied load. Tests are conducted at average walking speed (approximately 1 Hz), and often the loading cycle is a simplified gait cycle (Figure 2.2). Tests are generally short-term, with durations of between 300-1,200 cycles on average.

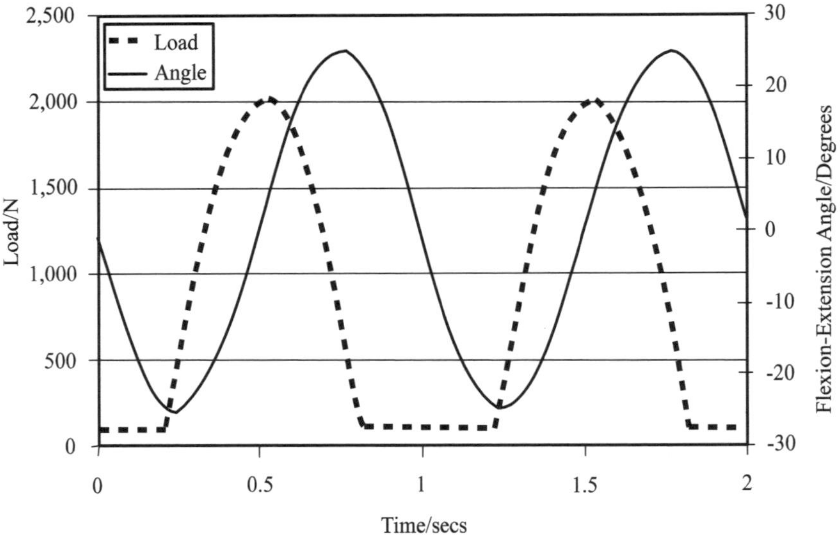

Figure 2.2. *Example input profile for a friction simulator study*

The effect of design, material, lubricant and other factors may be directly compared by contrasting the frictional torque and friction factor for each condition. Additionally, it is possible to assess the lubrication of the implant by conducting a Stribeck analysis. By altering one of the test conditions (lubricant viscosity, load or entraining velocity) it is possible to achieve a range of Sommerfeld numbers, which, when plotted against the friction factors calculated from the friction studies, can indicate the lubricating regime for the implant (Figure 2.3). A constant coefficient of friction (or friction factor) with a rising Sommerfeld number is indicative of boundary lubrication. Decreasing friction factor with an increasing Sommerfeld number indicates mixed lubrication, where an increasing proportion of the load is being supported by the lubricant, hence the friction is decreasing. Fluid film lubrication is demonstrated by an increasing friction factor with an increasing Sommerfeld number, caused by the viscous shear within the lubricant. Hence it is possible to determine, through testing a variety of experimental conditions, the lubricating regime of a total joint replacement. Most joints operate in a mixed or boundary regime, so there is a need for further study of the complex tribolayers that form on the surface of joints [WIM 03, SCH 01] and which may inhibit wear. Fluid lubrication, which may carry boundary layer molecules, plays an important role in enhancing these boundary layers.

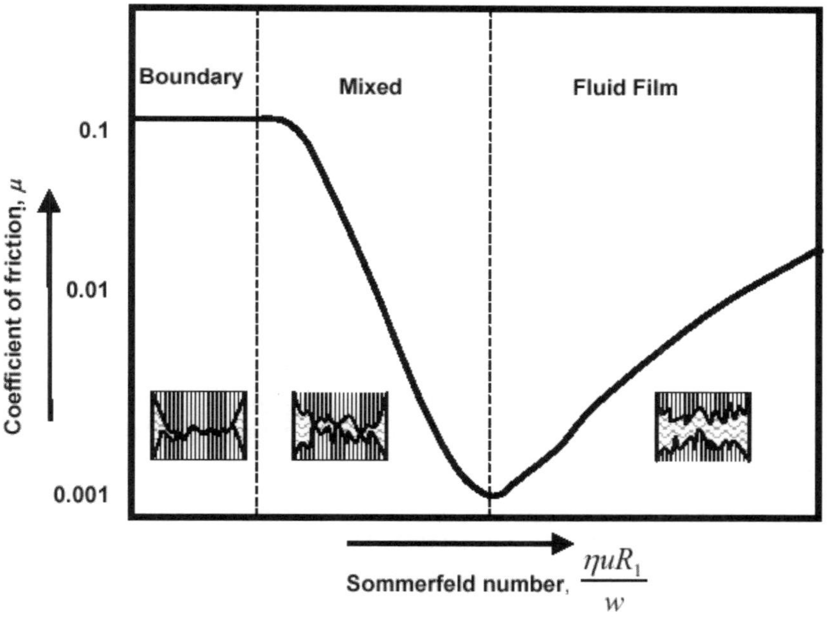

Figure 2.3. *Stribeck Analysis (figure copied from [UDO 03])*

2.2.3. *Wear*

Failure of total joint replacement may be attributed to mechanical wear and the response to wear particles [ING

00], therefore it is essential to assess the implant wear behavior in the laboratory to optimize implant design for longevity. There have been a broad variety of simulators and experimental methods for assessing wear, beginning with the simple configuration of pin-on-plate or pin-on-disk reciprocating rigs, which can assess material combinations and wear factors (Figure 2.4).

These are simple methods for evaluating new materials, and can also be used to generate wear debris for biological analysis. The load, rotation and sliding length can be adjusted in these studies to give slide tracks and contact pressures representative of those observed in a total joint replacement. Most importantly, they allow the investigation of effect of individual tribological parameters on wear. Assessment of the volumetric wear of the pin (W), through gravimetric analysis after a known period of cycles, and known total sliding distance (S), will enable calculation of the wear factor for the material, from equation [2.2] [GAL 06], where L is the compressive load applied.

$$K = \frac{W}{LS} \qquad [2.2]$$

The wear factor K in the above formulae is dependent on pressure. More recently an alternative wear formulation has been described for polyethylene such that volumetric wear (W) is dependent on sliding distance S and wear area A, with a dimensionless wear coefficient C, independent of load and pressure (equation [2.3], [FIS 10])

$$W = CAS \qquad [2.3]$$

In addition to these simple geometry test rigs, more advanced multi-axis wear simulators have also been developed. These allow assembled prostheses to be tested under physiological conditions, with the bearings orientated in an anatomical position, ensuring the loading vectors would be comparable to *in vivo* conditions.

Simulators have been developed for specific joint replacements, including knee (Figure 2.5), hip (Figure 2.6) and spine. They all have time-dependent loading, physiological motion and contain media (such as bovine calf serum) to recreate the environment for the replaced joint. Many of the simulators employ kinematic cycles derived from gait analysis such as the studies of the hip conducted by Paul *et al.* [PAU 66], and Lafortune *et al.* [LAF 92] in the knee. In addition, ISO standards exist which influence the test conditions used to study total joint replacements.

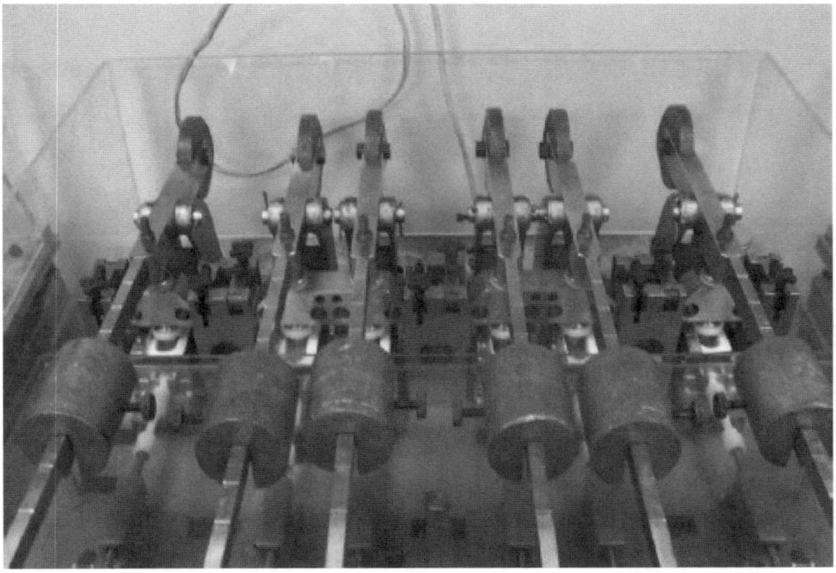

Figure 2.4. *Assembled pin-on-plate reciprocating rig*

During wear studies, a lubricant is used to replicate the *in vivo* conditions of synovial fluid within the joint. Bovine calf serum, diluted with water, is commonly used, as it has a similar protein composition to synovial fluid [COO 78, DOW

03], with concentrations being equivalent at approximately 25% bovine serum. However, the concentration, and indeed the media used within wear studies, varies greatly, and it should be noted that this may have a significant effect upon the wear outcome.

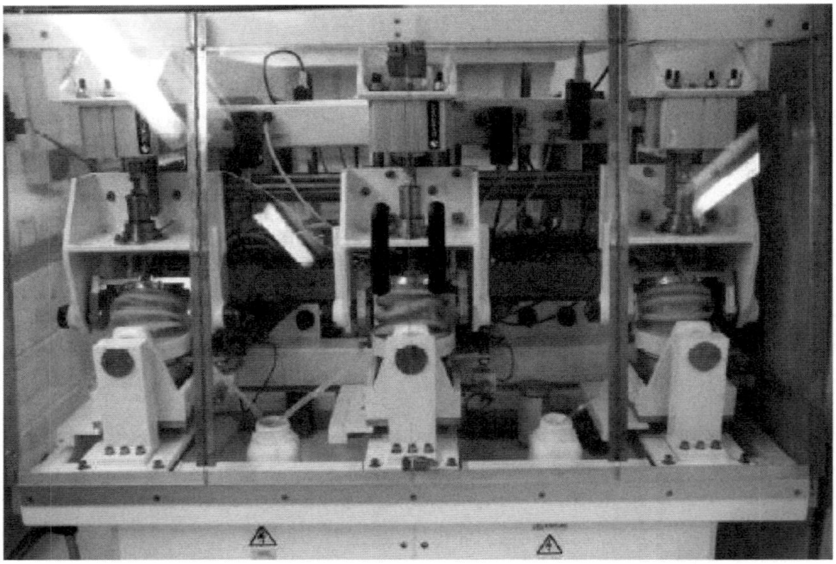

Figure 2.5. *Example knee simulator (Simulator Solutions, UK)*

Wear of components, either pins from the simple geometry studies, or implants, may be assessed by two different methods. The first is the most simple: the samples are removed from the simulator at given time intervals (for example, every one million cycles), cleaned and left to stabilize for a period of time (which varies with the material), after which the samples are weighed using a high accuracy/resolution balance. It is then possible to calculate the volumetric wear from the material properties. However, where the wear rates are very low, this method may be limited due to such factors as fluid uptake in polymeric materials, and material transfer.

An alternative method utilizes co-ordinate metrology, where the geometric changes in volume are measured by a contacting probe [BLU 08]. This also has limitations for low-wearing bearings, depending on the accuracy and resolution of the co-ordinate measuring machine (CMM).

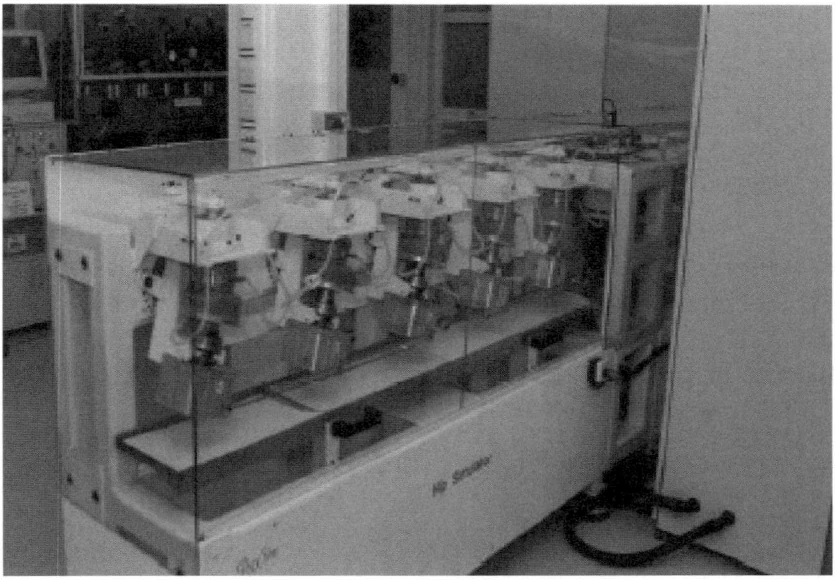

Figure 2.6. *Example of a hip simulator (Simulator Solutions, UK)*

Full component simulators may be used to identify the effects of material, design, kinematic input and other parameters upon the wear performance of a joint replacement, and may be useful in providing information in the design process. Most new materials undergo testing in the simple configuration pin-on-plate rig before full scale prostheses are made, to ensure the material has a satisfactory wear performance prior to manufacturing implants.

2.3. Effects of material and design on the tribology of total joint replacements

The design of an implant and the materials from which it is manufactured has a significant effect on the tribology of the joint. For example, if we review the Hamrock-Dowson equation for lubrication (equation [2.4]), applied to a total hip replacement, we can see that the minimum film thickness is dependent on load (W), entraining velocity (u), fluid viscosity (η), material properties (E), and design parameters of clearance and size (R). When calculating the lambda ratio (λ, equation [2.5]), which indicates the lubricating regime of the implant, surface roughness also has an effect (Ra). The most commonly implanted total joint replacements, and most widely studied experimentally, are the hip and knee replacement, and the following sections detail the effect of material and design parameters on the tribology of these bearings.

$$h_{min} = 2.789R\left\{\frac{\eta u}{E_{'}R}\right\}^{0.65}\left\{\frac{W}{E_{'}R^2}\right\}^{-0.21} \qquad [2.4]$$

$$\lambda = \frac{h_{min}}{\left[(Ra_1)^2 + (Ra_2)^2\right]^{1/2}} \qquad [2.5]$$

2.3.1. *Total hip and resurfacing replacements*

The effect of material and design on the tribology of total hip replacements, and large diameter surface replacements has been studied extensively. This section examines the effect of material combination on the wear and friction of total joint replacement, and the effect of design changes, such as bearing diameter and clearance, on the tribology of the hip replacement for one material combination.

2.3.1.1. *Effect of material combination on the tribology of total hip replacements*

2.3.1.1.1. Friction

The tribology of total hip replacements may be influenced by the material combination of the implant. If we review equations [2.4] and [2.5], which assist in theoretically predicting the tribology of the implant, we can see the material has an influence in two respects, the mechanical properties of the material and the surface finish. For a standard 28 mm total hip replacement, the lambda ratios calculated for the materials, under a load of 2 kN, and an entraining velocity based on a flexion-extension motion of $\pm$ 25°/s, predict different lubricating regimes and highlight the differences in tribology of the total hip replacements (Table 2.1, [JIN 97]).

Bearing combination	**λ ratio**	**Lubricating regime**
Metal-on-metal	< 1	Boundary
Metal-on-polyethylene	< 1	Boundary
Ceramic-on-polyethylene	< 1	Boundary
Ceramic-on-metal	> 3	Fluid film
Ceramic-on-ceramic	> 3	Fluid film

Table 2.1. *Predicted lubricating regimes for total hip replacement*

Several friction studies have been conducted which examine the influence of material on the friction of the total hip replacement bearing. Comparing materials under identical test conditions, ceramic-on-ceramic bearings have been shown to have the lowest friction factors, with values between 0.04-0.06 [BRO 07, 08; SCH 00a, 01, 06], the frictional torque associated with ceramic-on-ceramic bearings has also been identified as being the lowest compared with other standard bearing combinations [BIS

08]. Ceramic-on-metal bearings have similar frictional torque and friction factor to the ceramic-on-ceramic bearings [BRO 07, BIS 08]. The highest friction has been measured in metallic bearings, when testing equivalent geometry and size (approximate friction factors of 0.12-0.18 when lubricated in 25% bovine serum).

Studies of dry friction for metal-on-metal contact has demonstrated higher friction than predicted by simple friction theory, with the experimental friction coefficient for a chromium-chromium contact to be approximately 0.4 [DOW 81]. Due to material similarities, it is possible for adhesive junctions to form where asperities meet in metal-metal contacts, thus the increased shear stress required to break these contacts results in elevated friction. As the metal-on-metal hip implant operates within a boundary to mixed lubrication regime throughout a test cycle, there is potential for these adhesive joints to form through asperity contact. By changing from a metal-metal contact to a ceramic-metal contact in a ceramic-on-metal bearing, there is no possibility of these junctions forming and hence the friction is significantly reduced.

The softer polymeric materials may benefit from an enhanced fluid layer due to a lower Young's modulus compared with the hard-on-hard materials, however, an increased surface roughness of the UHMWPE bearings compared with metal or ceramic means this advantage is limited as the asperities would penetrate the lubricating film. However, these materials may potentially undergo greater deformation compared to the hard-on-hard bearings, hence exhibiting lower friction than the metal-on-metal bearings. Friction factors for metal or ceramic bearings coupled with UHMWPE tend to be higher than ceramic-on-ceramic or ceramic-on-metal bearings (range of 0.05-0.65), but not significantly so. Two studies have found metal-on-polyethylene bearings to have higher friction factors and frictional torque than alumina ceramic-on-polyethylene, but this has not been shown to be significant [BRO 07, SAI 09].

One study has utilized synthetic thickening agents to modify the viscosity of the lubricant, to allow Stribeck analysis to be performed on the different materials. This study showed that the ceramic-on-ceramic bearing operates within a fluid film lubrication regime during the test cycle, but the metal-on-metal and polyethylene bearings operate in a mixed lubrication regime [SCH 06]. However, when using a biological lubricant, all combinations are within the mixed lubricating regime [BRO 07, SCH 00a].

The lubricant used within experimental tribological studies has a significant effect upon the friction of the bearing. Previously, it was noted that bovine serum was considered an appropriate lubricant, as it has similar protein composition compared to synovial fluid. However, variations in concentration, and studies with other fluid media, have shown these proteins to have a significant effect upon the friction of bearings.

An increasing protein concentration in a metal-on-metal study, such as moving from a study in water through to 100% bovine serum, will result in a reduction in friction. Proteins may be adsorbed onto each of the metallic surfaces, acting as a boundary lubricant to reduce the potential for asperity adhesion, hence reducing the friction within the bearing. It has been shown that boundary lubrication films play a significant role in both friction reduction and low wear in metal-on-metal bearings [SCH 00a, WIM 03]. Conversely, all other bearings exhibit increased friction with increasing protein concentration, due to the additional forces required to shear the proteins within the fluid media.

Other studies have been conducted to investigate the impact of other biological media on the friction of bearings (specifically in metal-on-metal joints), and have shown mixed outcomes. One study suggests the larger macro-molecules in blood contribute to the increased friction observed in the study [MCM 06].

2.3.1.1.2. Lubrication

One study has used electrical resistance methodology to assess the lubrication of metal-on-metal and ceramic-on-ceramic bearings under wear simulator conditions. Due to the electrical properties of polyethylene, it was not possible to study the lubrication via this method, and the ceramic bearings were coated to allow sufficient conductivity [SMI 01]. Both bearings demonstrated separation during the swing phase of the gait cycle, but the ceramic bearings appeared to have longer periods of separation compared to the metal-on-metal bearings.

2.3.1.1.3. Wear

A broad range of wear simulator studies have been undertaken to examine the wear of different materials. These have indicated that hard-on-hard bearings tend to have a lower volumetric wear rate than hard-on-soft bearings for a comparable bearing size. Furthermore, these bearing types appear to exhibit a different wear trend with time [AHL 97, CHA 99, CLA 00, GOL 00, DOW 01, ESS 05]. Metal or ceramic-on-polymer bearings tend, after an initial phase of creep, to have a reasonably steady wear rate throughout the duration of the wear tests, whereas hard-on-hard bearings tend to have a short period of relatively high wear, called the *bedding-in period*, tending towards a steady-state low wear rate [CAM 99, ANI 01, SCH 01, LES 08, HU 04].

At present, most of the total hip replacements implanted around the world still tend to be either a ceramic or metal femoral head articulating with a polyethylene, or cross-linked polyethylene acetabular cup. Conventional polyethylene, articulating with ceramic or metallic heads has been shown to have wear rates between 25-40 mm^3 per million cycles of testing, with a relatively linear wear performance over time [BIG 97, WIL 03, ESS 05, GAL 06].

Clinically, failure of total hip replacements is rarely related to the bearing "wearing-out", however, wear of an implant causes wear debris to be released into the surrounding tissue. This particulate debris may build up in the tissues, until a sufficient volume is achieved which causes an immune response. The total wear volume required to generate an osteolytic response lies between 300-700 mm^3. This may result in osteolysis and loosening of the total hip replacement, causing failure of the implant [ING 00, 05]. Therefore it is advantageous to reduce the volumetric wear of polyethylene.

Cross-linking the polyethylene material, through exposure to gamma irradiation or an electron beam, can increase wear resistance by creating links between the polymer chains. Wear studies of cross-linked polyethylene have shown a significant reduction in wear compared with conventional polyethylene bearings, when articulated with either ceramic or metal heads.

The level of cross-linkage in a polyethylene bearing is determined by the level of radiation, and is defined as either moderately cross-linked (approximately 5 MRad) or highly cross-linked (approximately 10 MRad). The wear rate of cross-linked polyethylene significantly reduces with increasing radiation level [GAL 06]. It is clear that a reduced wear rate associated with the cross-linked polyethylene will reduce the potential for osteolysis. *In vitro* wear studies have also indicated a lower wear rate for polyethylene tested with ceramic heads compared with metallic heads [WIL 03, GAL 06].

Hard-on-hard bearings provide an alternative option for reducing volumetric wear and wear debris. Ceramic-on-ceramic bearings (both alumina and zirconia-toughened alumina (ZTA)) have shown very low wear rates under standard simulator conditions [NEV 00, NEV 01], and at elevated cup angles (up to 60° inclination), with less than 0.5 mm^3 of wear per million cycles of test.

However, a different wear phenomenon has been observed *in vivo* where a stripe region of wear was noted on the head, which corresponded to a region of wear on the cup at the rim. Analysis of retrieved ceramic-on-ceramic bearings showed that during the gait cycle, the head may displace from the cup by a small amount, through the swing phase, and upon heel strike, the head would impact upon the edge of the cup, resulting in the stripe wear observed.

An experimental method (micro-separation) was devised to recreate these conditions in a wear simulator, by displacement of the bearings, relative to one-another during the swing phase load of the cycle [NEV 00, STE 03]. This resulted in a significant increase in wear for the alumina bearings, compared to a standard condition study, and corresponded well with clinical data, both in terms of wear rate and location of the wear stripe (Figure 2.7). The experimental study demonstrated that a ZTA-ZTA bearing was more resistant to this mode of wear than the alumina-alumina combination [STE 03].

In addition to showing a significant reduction in wear compared to the conventional and cross-linked polyethylene bearings during hip simulator studies, the biological activity of the ceramic wear debris is very low compared with polyethylene, and therefore this debris is unlikely to cause failure through osteolysis [GER 03].

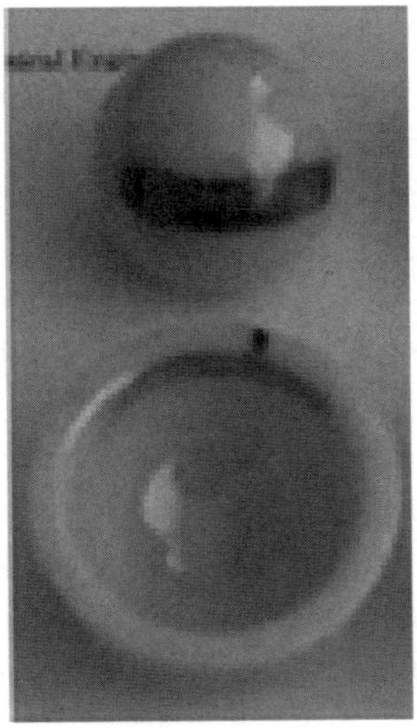

Figure 2.7. *Image demonstrating stripe wear on an alumina head and cup (this image is achieved by lead-pencil rubbing on the wear regions; [STE 03])*

Numerous wear studies have been conducted to investigate the wear performance and lubrication in metal-on-metal total hip replacements. A self-polishing phenomenon has been noted during studies, where the surface roughness has progressively improved, reducing asperity contact and enhancing wear performance. The wear behavior of the metal-on-metal bearing is noted for a relatively high bedding-in phase of wear, followed by a low wearing steady-state phase, with the transition period occurring within the first million cycles [CLA 00].

It has been proposed that during the bedding-in phase of the wear, the metallic bearings are tending towards optimal conformity to achieve the best tribological conditions; hence this is followed by very low wear.

Mean wear rates reported for metal-metal bearings depend on the composition of the metal and the lubricant used within the study; however, many studies indicate steady-state wear rates below 1 mm^3 per million cycles for conventional 28 mm bearings, a significant reduction compared with the polyethylene bearings [SCH 01, FIR 01a]. The volume of debris generated by metal-metal bearings is significantly lower than polyethylene bearings. However, the particle size is reduced and therefore the number of particles may be higher [DOO 96, CAT 03, ING 05].

Although modern generation metal-metal bearings are not associated with osteolysis, the small particles of metal debris are able to enter cells, and patients with metal-metal bearings may exhibit elevated metal ion levels in the blood [HAR 08, WIL 96, KEE 07]. The long-term effect of metal ions within the body is not yet understood.

Analysis of retrieved bearings, clinical studies and experimental studies have all suggested that elevated metal ion levels may be caused by rim loading, where the head is articulating, at least through part of the gait cycle, with the edge of the cup, causing high wear [LES 09, WIL 08]. The novel differential hardness ceramic-on-metal bearing has shown similar wear rates to ceramic-on-ceramic bearings *in vitro* [FIR 01b, WIL 08, ISA 09].

The wear rates of ceramic-on-metal bearings have been shown to be significantly lower than metal-on-metal bearings. Due to the dissimilar materials, there is reduced adhesive wear, the smoother ceramic bearing may reduce abrasive wear, and corrosive wear is also reduced. Under micro-separation conditions, there was no evidence of stripe wear on the ceramic heads, with only slight evidence of metallic transfer from the cups [BRO 07]. Clinical studies have shown a significant reduction in metal ion levels in the blood compared with metal-on-metal bearings [ISA 09].

It is clear that material has a significant impact upon the tribology and wear of a total hip replacement. However, other design factors, such as clearance and diameter also have a significant effect on the tribology of the bearing. It is interesting to note that the impact of these factors varies with the material, and therefore it is important to consider the interaction of the material and design parameter when reviewing the tribology of the bearing.

2.3.1.2. *Effect of bearing diameter on the tribology of total hip replacements*

Under fluid-film lubrication conditions, theory would predict an increase in bearing diameter, which would enhance film thickness and therefore improve the tribology of larger diameter bearings. However, under boundary lubrication, an increase in bearing diameter results in an increase in sliding distance and therefore increased wear.

A study by Smith *et al.* [SMI 01] demonstrated that for small diameter metal-on-metal bearings, the ratio of wear was equivalent to head size, indicating the increased bearing size resulted in increased sliding distance and thus wear. However, at diameters greater than 28 mm this ratio no longer applied, and the wear rates were reduced [DOW 04]. Electrical resistance techniques indicated increasing separation, with a fluid-film lubrication evident during phases of the gait cycle. More recent studies involving metallic surface replacements have also indicated reducing wear with increasing bearing size, indicating better tribological conditions [LES 08].

O'Connor *et al.* demonstrated that femoral head size had an influence on friction using a pendulum comparator [OCO 98], and studies since then have shown a reducing friction factor with increasing head diameter for metal-on-metal bearings [BRO 08].

Studies examining the influence of bearing diameter in polyethylene bearings have shown an increase in wear with increasing bearing diameter [FIS 91, JEN 07]. As polyethylene bearings tend to operate in the boundary lubrication regime, the larger bearing diameter results in increased sliding distance and hence elevated wear. Thus, clinically, it is rare to see total hip replacements using a polyethylene bearing with a diameter in excess of 36 mm. If we review the formulae for polyethylene wear (equation [2.3]), a doubling in head size effectively results in an eight-fold increase in wear, hence little osteolysis is seen in the original Charnley bearings of 22 mm head size, compared with the larger bearings [FIS 10].

2.3.1.3. *Effect of bearing clearance on the tribology of total hip replacements*

In addition to the diameter, bearing clearance also has a significant influence on the tribology of a hip replacement, and again, the trends displayed depend on the material combination for the hip replacement. Clearance can be defined either as radial or diametral, and is the difference between the head diameter and the cup diameter. Reduced clearances in metal-on-metal bearings have been shown to result in enhanced tribology. When metallic bearings of comparable diameter were studied, a reduction in wear was observed with reducing clearance [SMI 01, DOW 04, RIE 05, HU 04]. However, at small clearance, manufacturing precision becomes critical, as deviation in sphericity or tolerance may cause the joint to seize, increasing both wear and friction [HU 06]. A study examining friction in large diameter surface replacements showed a similar reduction in friction with reducing clearance [BRO 08].

Experimental assessment of lubrication has also reviewed the effect of clearance. Both resistance and ultrasonic measurement of bearing separation have indicated greater levels of separation with reducing clearance. An ultrasonic assessment during friction studies showed increasing film thickness with reduced clearance [BRO 08]. Two resistance studies conducted in a hip wear simulator have also indicated longer periods of separation during the gait cycle with reduced clearance, suggesting that the lower clearance bearings operate within fluid film lubrication for longer in each gait cycle than the larger clearance bearings [MCN 00, SMI 01].

Conversely, studies in metal-on-polymer bearings have indicated there is a critical minimum clearance, below which wear will increase. A reduced clearance may affect the ability of a joint to be lubricated. When the clearance becomes small, areas may become depleted of lubrication which would cause greater contact between the two bearing surfaces, resulting in increased wear [FIS 91].

In addition to the material, design considerations such as bearing diameter and clearance have a significant effect on the tribology of a hip replacement. In addition to these factors, surgical considerations such as cup inclination angle, head offset and cup deformation, as well as patient factors, such as activity and weight, will also have a significant impact on the tribology of the hip replacement in the clinical environment.

2.3.2. *Total knee replacement*

The total knee replacement is a more complex geometry than a total hip replacement, and therefore other factors may contribute to the performance of a total knee replacement. Virtually all knee replacements are comprised of metallic femoral and tibial bearings, with polyethylene inserts; therefore boundary lubrication is the dominant lubricating regime in the knee. Due to the materials employed, it is clear that it is not possible to experimentally evaluate the lubrication in the knee through the ultrasonic and electrical resistance techniques previously outlined, therefore the assessment of the tribology of the knee is conducted through wear studies. Due to the complex kinematics and geometry, other factors may influence the wear behavior of the knee, which will be considered in the next few sections.

2.3.2.1. *Influence of bearing design on wear in total knee replacement*

The large range of total knee replacements commercially available may be essentially categorized based on whether the bearing insert is fixed or mobile. A fixed bearing knee will have a polyethylene insert which will be fixed into a tibial tray so that very little or no motion occurs between the insert and the tray. This means that the complex motion of the knee, including the flexion-extension and internal-external rotation occur on the upper surface of the insert, between the femoral component and the insert (Figure 2.8).

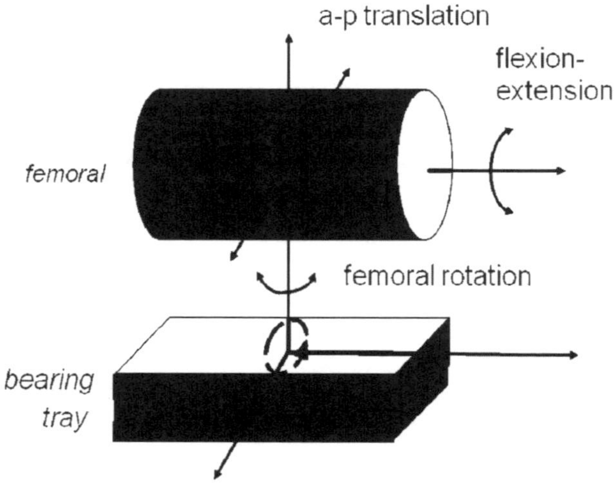

Figure 2.8. *Schematic of a fixed bearing knee design*

Alternatively, a mobile bearing knee allows movement between the tibial tray and the insert. This group of knees may be sub-categorized into bearings which allow anterior-posterior translation and rotation to occur on the lower surface, and those which only permit rotation to occur between the tibial tray and insert, called "rotating platform" bearings. The rotating platform design essentially translates the complex motion of the knee into two simple uni-directional motions at the two interfaces of the bearing insert. Linear motion parallel to the axis of flexion-extension occurs on the upper surface of the insert, and uni-directional rotation occurs on the lower surface ([MCE 05] Figure 2.9).

In a fixed bearing knee, the multi-directional motion causes increased cross-shear. A material property of polyethylene is that strain hardens along the principal direction sliding, due to molecular alignment. However, although wear resistance is increased in this direction, it is weakened in the transverse direction, and therefore increased cross-shear on the upper surface of the bearing insert will result in increased wear. The decoupled motion that occurs in a rotating platform mobile bearing knee reduces cross-shear and hence results in lower wear [MCE 05, UTZ 09, GRU 09].

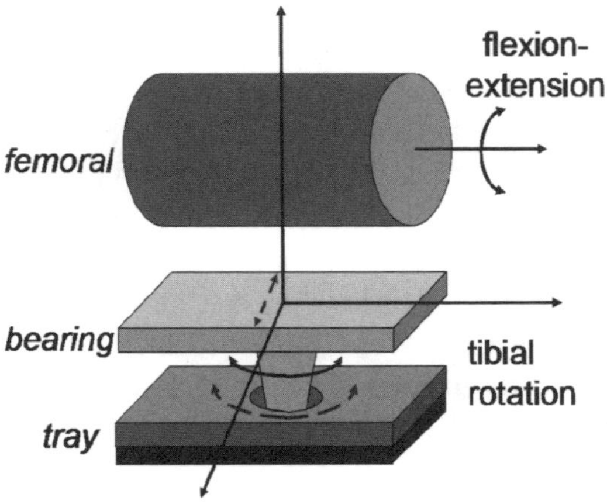

Figure 2.9. *Schematic of a rotating platform mobile bearing knee replacement*

2.3.2.2. *Influence of bearing material on wear in total knee replacement*

Failure of early designs of total knee replacement could often be attributed to delamination and fatigue failure of the polyethylene bearing. Historically, UHMWPE was gamma sterilized in air; sterilization in the presence of oxygen allowed the free radicals generated by the gamma irradiation to bind with the oxygen, resulting in degradation and a significant reduction in mechanical properties. Current polyethylene materials are either sterilized in a vacuum or an inert environment to prevent this binding occurring, and hence retain material properties. Simulator studies have shown a significant improvement in wear behavior, and no incidence of delamination when comparing the current polyethylene with historical aged polyethylene [YOU 00].

The issues highlighted in the previous sections regarding polyethylene wear also apply to a knee bearing design, and therefore cross-linked polyethylene has become increasingly popular. However, increasing cross-linkage results in a reduction in mechanical toughness of the material, and thus there is a balance between wear resistance and fatigue resistance. It is common for the polyethylene in total knee replacement to be moderately cross-linked (5 MRad), compared to the highly cross-linked material used in the hip. Experimental wear studies have shown a significant reduction in volumetric wear with cross-linked polyethylene compared to conventional polyethylene in total knee replacement [FIS 04, FIS 10].

In the hip, we saw an advantage of ceramic-on-polyethylene compared with metal-on-polyethylene due to reduced wear and greater resistance to scratching and damage by third body debris. However, concerns remain regarding the fracture toughness of the material for applications to complex and potentially thin geometries such as femoral bearing. A small number of studies examining ceramic femoral components articulating with polyethylene compared to a cobalt-chromium bearing showed a significant reduction in wear rate under the same conditions [LEE 09, OON 09]. However, further research and clinical studies are required to confirm this potential benefit.

2.3.2.3. *Effect of conformity on wear in knee replacement*

In the last section, we diskussed the failure of total knee replacement through delamination of the polyethylene insert. Delamination may occur through poor material properties, but may also be related to high stresses in the material. One design approach to reduce the incidence of failure in the knee replacement was to modify the design to reduce the contact stresses. Often knee replacement designs have a moderately conforming contact between the femoral bearing and the polyethylene insert, which reduces contact stress and increases the area of contact. The improvements in mechanical properties of polyethylene, through enhanced material processing have enabled the effect of conformity on the wear of total knee replacements to be investigated [GAL 09, FIS 10].

Several studies have shown a reduction in wear with a lower conformity bearing, however, a study contrasting a curved and a completely flat insert, in a fixed bearing design knee, under identical test conditions showed a significant reduction in wear [GAL 09]. Due to a reduction in conformity, the contact area is reduced.

Throughout the knee articulation, the area of contact is significantly smaller than in a conforming bearing, and hence the potential for surface wear is reduced. However, the contact pressures in the bearing are increased, and therefore it is only due to the improvements in mechanical properties of the material that this design change may be considered. A study contrasting a low conformity partial knee replacement with a highly conforming spherical knee replacement has shown a similar advantage, with small wear areas and significantly lower wear rates for the reduced conformity bearing [BRO 10].

2.4. Conclusion

Clearly there are many aspects that may influence the biotribology of a total joint replacement, and this chapter has touched on only the two most common replacement joints. As technology extends into other bearings, such as ankles and spinal disk replacement, the tribological principals outlined within this chapter must be considered in these designs. In addition to exploring the influence of material and bearing design experimentally, we must also consider new factors which present themselves during surgery, and throughout the patient's lifetime, such as bearing alignment, soft tissue balancing and patient activity. At present, experimental studies often examine well aligned bearings under a normalized loading and motion cycle, which could be considered somewhat of an idealized test condition.

In the future, to differentiate between the performance of new designs and to more closely replicate clinical outcomes, sub-optimal conditions should be investigated, such as purposefully setting up mis-aligned bearings, causing damage to the bearing surfaces, and increasing the loading and motion cycles to be more reflective of typical patient activity. Experimental studies under standard gait conditions have enabled the prediction of performance of total joint replacement under good conditions. These are not the conditions that cause bearings to fail clinically, therefore the future drive for experimental studies of total joint replacement must be towards adverse conditions, and new bearings need a more robust tribological design which can tolerate these adverse conditions.

2.5. Bibliography

[AHL 97] AHLROOS T., SAIKKO V., "Wear of prosthetic joint materials in various lubricants", *Wear*, vol. 211 no. 1, pp. 113-119, 1997.

[ANI 01] ANISSIAN H.L., STARK A., GOOD V., DAHLSTRAND H, CLARKE IC., "The wear pattern in metal-on-metal hip prostheses", *Journal of Biomedical Materials Research*, vol. 58 no. 6, pp. 673-678. 2001.

[BIG 97] BIGSBY R.J.A., HARDAKER C., FISHER J., "Wear of ultra-high molecular weight polyethylene acetabular cups in a physiological hip joint simulator in the anatomical position using bovine serum as a lubricant", *Proceedings of the Institution of Mechanical Engineers, Part H, Journal of Engineering in Medicine*, vol. 211, 1997.

[BIS 08] BISHOP N., WALDOW F., MORLOCK M.M., "Friction moments of large metal-on-metal hip joint bearings and other modern designs", *Medical Engineering and Physics*, 2008.

[BLU 08] BLUNT L.A., BILLS P.J., JIANG X.-Q., CHAKRABATY G "Improvement in the assessment of wear of total knee replacements using coordinate-measuring machine techniques", Proceedings *of the Institution of Mechanical Engineers. Part H, Journal of Engineering in Medicine*, pp. 222, 2008.

[BRO 07] BROCKETT C., WILLIAMS S., JIN Z., ISAAC G., FISHER J., "Friction of total hip replacements with different bearings and loading conditions", *Journal of Biomedical Materials Research Part B*, 2007

[BRO 08] BROCKETT C.L., HARPER P., WILLIAMS S., ISAAC G.H., DWYER-JOYCE R.S., JIN Z., FISHER J., "The influence of clearance on friction, lubrication and squeaking in large diameter metal-on-metal hip replacements", *J. Mater. Sci: Mater. Med.*, vol. 19, pp. 1575–1579, 2008

[BRO 10] BROCKETT C.L., JENNINGS L.M., FISHER J., "A comparison of the wear of fixed and mobile bearing unicompartmental knee replacements", *14th Congress of the European Society of Sports Traumatology, Knee Surgery and Arthroscopy*, ESSKA, 2010

[CAM 99] CAMPBELL P., MCKELLOP H, ALIM R, MIRRA J, NUTT S, DORR L, AMSTUTZ HC. "Metal-on-metal hip replacements: Wear performance and cellular response to wear particles", *ASTM Special Technical Publication*, vol. 1365, pp. 193-209, 1999.

[CAT 03] CATELAS I., BOBYN J.D., MEDLEY J.B., KRYGIER J.J., ZUKOR D.J., HUK O.L., "Size, shape, and composition of wear particles from metal-metal hip simulator testing: effects of alloy and number of loading cycles", *Journal of Biomedical Materials Research - Part A*, vol. 67 no. 1, pp. 312-327, 2003.

[CHA 99] CHAN F.W., BOBYN J.D., MEDLEY J.B., KRYGIER J.J., TANZER M., "The Otto Aufranc Award - Wear and lubrication of metal-on-metal hip implants", *Clinical Orthopaedics and Related Research*, vol. 369, pp. 10-24, 1999.

[CHA 66] CHARNLEY J., "An artificial bearing in the hip joint: implications in biological lubrication", *Federation Proceedings*, vol. 25 no. 3, pp.1079-81, 1966.

[CLA 00] CLARKE I.C., GOOD V., WILLIAMS P., GUSTAFSON G., SCHROEDER D., ANISSIAN L., STARK A., OONISHI H., J. S., "Ultra-low wear rates for rigid-on-rigid bearings in total hip replacements", *Proceedings of the I MECH E Part H Journal of Engineering in Medicine*, pp. 331-347, 2000.

[COO 78] COOKE A.F., DOWSON D., WRIGHT V., "The rheology of synovial fluid and some potential synthetic lubricants for degenerate synovial joints", *Engineering in Medicine*, vol. 7, pp. 66-72. 1978.

[DOO 96] DOORN P.F., MIRRA J.M., CAMPBELL P.A., AMSTUTZ H.C., "Tissue reaction to metal on metal total hip prostheses", *Clinical Orthopaedics & Related Research*, vol. 329 Suppl. S187-205, 1996.

[DOW 81] DOWSON D., UNSWORTH A., COOKE A.F., GVOZDANOVIC D., "Lubrication of joints", in: DOWSON D, WRIGHT V (Eds.), *An Introduction to the Bio-mechanics of Joints and Joint Replacements*, Mechanical Engineering Publication Ltd. London, pp. 120-145, 1981.

[DOW 01] DOWSON D., "New joints for the Millennium: wear control in total replacement hip joints", *Proceedings of the Institution of Mechanical Engineers*, Part H, *Journal of Engineering in Medicine*, vol. 215, no. 4, pp. 335-358, 2001.

[DOW 04] DOWSON D., HARDAKER C., FLETT M.S., ISAAC, G.H., "A hip joint simulator study of the performance of metal-on-metal joints: Part II: Design", *The Journal of Arthroplasty*, 19(8, Supplement 1):124-130, 2004.

[ESS 05] ESSNER A., SUTTON K., WANG A., "Hip simulator wear comparison of metal-on-metal, ceramic-on-ceramic and crosslinked UHMWPE bearings", *Wear*, vol. 259, no. 7-12, pp. 992-995, 2005.

[FIR 01a] FIRKINS P.J., TIPPER J.L., INGHAM E., STONE M.H., FARRAR R., FISHER J., "A novel low wearing differential hardness, ceramic-on-metal hip joint prosthesis", *Journal of Biomechanics*, vol. 34, no. 10, pp. 1291-1298, 2001.

[FIR 01b] FIRKINS PJ, TIPPER JL, INGHAM E, STONE MH, FARRAR R, FISHER J., "Influence of simulator kinematics on the wear of metal-on-metal hip prostheses", *Proceedings of the Institution of Mechanical Engineers, Part H, Journal of Engineering in Medicine*, vol. 215 no. 1, pp. 119-122, 2001.

[FIS 91] FISHER J., DOWSON D., "Tribology of total artificial joints", *Proceedings of the Institution of Mechanical Engineers*, Part H, *Journal of Engineering in Medicine*, vol. 205 no. 2, pp. 73-9, 1991.

[FIS 04] FISHER J., MCEWEN H.M.J, TIPPER J.L. *et al.*, "Debris and biological activity of cross-linked polyethylene in the knee CORR", *Wear*, vol. 428, pp. 114-119, 2004.

[FIS 10] FISHER J., JENNINGS L.M., GALVIN A.L., JIN Z., STONE M., INGHAM E., "Polyethylene wear in total knees 2009", Knee Society Presidential Guest Lecture, *Clinical Orthopaedics and Related Research*, vol. 468, 2010.

[GAL 06] GALVIN A., KANG L., TIPPER J., STONE M., INGHAM E., JIN Z., FISHER J., "Wear of cross-linked polyethylene under different tribological conditions", *J. Mater. Sci: Mater. Med.*, vol. 17, pp. 235-243, 2006.

[GAL 09] GALVIN A.L., KANG L., UDOFIA I., JENNINGS L.M., FISHER J., "Effect of conformity and contact stress on wear in fixed bearing total knee prostheses", *J. Biomechanics*, vol. 42, pp. 1898-1902, 2009.

[GER 03] GERMAIN M.A., HATTON A., WILLIAMS S., MATTHEWS J.B., STONE M., FISHER J., INGHAM E., "Comparison of the cytotoxicity of clinically relevant cobalt–chromium and alumina ceramic wear particles in vitro", *Biomaterials*, vol. 24, 2003.

[GOL 00] GOLDSMITH A.A., DOWSON D., ISAAC G.H., LANCASTER J.G., "A comparative joint simulator study of the wear of metal-on-metal and alternative material combinations in hip replacements", *Proceedings of the Institution of Mechanical Engineers, Part H, Journal of Engineering in Medicine*, vol. 214 no. 1, pp. 39-47, 2000.

[GRU 09] GRUPP T.M., KADDICK C., SCHWIESAU J. *et al.*, "Fixed and mobile bearing total knee arthroplasty – influence on wear generation, corresponding wear areas, knee kinematics and particle composition", *Clinical Biomechanics*, vol. 24, pp. 210-217, 2009.

[HAR 08] HART A.J., BUDDHDEV P., WINSHIP P., FARIA N., POWELL J.J., SKINNER J.A., "Cup inclination angle of greater than 50 degrees increases whole blood concentrations of cobalt and chromium ions after metal-on-metal hip resurfacing", *Hip International*, 2008.

[HU 04] HU X., ISAAC G., FISHER J., "Changes in the contact area during the bedding-in wear of different sizes of metal on metal hip prostheses", *Biomedical Materials and Engineering*, vol. 14, 2004.

[HU 06] HU X.Q., TWIGG P., SHELTON J.C., TUKE M., TAYLOR A., "Frictional study of metal on metal hip prostheses with different clearances", *52nd Annual Meeting of the Orthopaedic Research Society*, Chicago, USA, Paper 0502, 2006.

[ING 00] INGHAM E., FISHER J., "Biological reactions to wear debris in total joint replacement", *Proceedings of the Institution of Mechanical Engineers, Part H, Journal of Engineering in Medicine*, vol. 214, no. H1, pp. 21-37, 2000.

[ING 05] INGHAM E., FISHER J., "The role of macrophages in osteolysis of total joint replacement", *Biomaterials*, vol. 26 no. 11, pp. 1271-1286, 2005.

[ISA 09] ISAAC G., BROCKETT C., BRECKON A., VAN DER JAGT D., WILLIAMS S., HARDAKER C., FISHER J., SCHEPERS A., "Ceramic-on-metal bearings in total hip replacement", *JBJS*, vol. 91B, 2009.

[JIN 97] JIN Z.M., DOWSON D., FISHER J., "Analysis of fluid film lubrication in artificial hip joint replacements with surfaces of high elastic modulus", *Proceedings of the Institution of Mechanical Engineers, Part H, Journal of Engineering in Medicine*, vol. 211 no. 3, pp. 247-56, 1997.

[KEE 07] KEEGAN G.M., LEARMONTH I.D., CASE C.P., "Orthopaedic metals and their potential toxicity in the arthroplasty patient: A review of current knowledge and future strategies", *J. Bone Joint Surg. [Br],* vol. 89-B, pp. 567-573, 2007.

[JEN 07] JENNINGS L.M., GALVIN A.L., FISHER J., *Wear and Creep of Highly Cross Linked Polyethylene Against Cobalt Chrome and Ceramic Femoral Heads*, British Hip Society, Leeds, UK, 1-2 March 2007.

[LAF 92] LAFORTUNE M.A., CAVANAGH P.R., SOMMER III H.J., KALENAK A., "Three-dimensional kinematics of the human knee during walking", *J. Biomechanics*, vol. 25, pp. 347-357, 1992.

[LEE 09] LEE J.K.L., MARUTHAINAR K., WARDLE N., "Increased force simulator wear testing of a zirconium oxide total knee arthroplasty", *The Knee*, vol. 16, p 269-274, 2009.

[LES 08] LESLIE I.J., WILLIAMS S., BROWN C., ISAAC G., JIN Z., INGHAM E., FISHER J., "Effect of bearing size on the long-term wear, wear debris, and ion levels of large diameter metal-on-metal hip replacements – an *in vitro* study", *J. BMR*, Part B, 2008.

[LES 09] LESLIE I.J., WILLIAMS S., ISAAC G., INGHAM E., FISHER J., *High Cup Angle and Microseparation Increase the Wear of Hip Surface Replacements*, CORR, 2009.

[MCE 05] McEWEN H.M.J., BARNETT P.I., BELL C.J., FISHER J., "The influence of design, materials and kinematics on the in-vitro wear of total knee replacements", *J. Biomechanics*, vol. 38, pp. 357-365, 2005.

[MCM 06] McMINN D., DANIEL J., KAMALI A., SAYAD SARAVI S., YOUSEFFI M., BAND T., ASHTON R., "Friction testing in metal-metal bearings with different clearances using blood as lubricant", *52nd Annual Meeting of the Orthopaedic Research Society*, Chicago, USA, 2006.

[MCN 00] McNIE C.M., DOWSON D., "The effect of radial clearance on lubrication in a metal on metal joint tested in a hip joint simulator. Thinning films and tribological interfaces", *Tribology and Interface Engineering Series*, vol. 38, pp. 341-345, 2000.

[NEV 00] NEVELOS J.E., INGHAM E., DOYLE C., STREICHER R.M., NEVELOS A.B., WALTER W., FISHER J., "Microseparation of the centres of alumina-alumina artificial hip joints during simulator testing produces clinically relevant wear rates and patterns", *J. Arthroplasty*, vol. 15, pp. 793-795, 2000.

[NEV 01] NEVELOS J.E., INGHAM E., DOYLE C., NEVELOS A.B., FISHER J., "The influence of acetabular cup angle on the wear of 'BIOLOX Forte' alumina ceramic bearing couples in a hip joint simulator", *Journal of Materials Science: Materials in Medicine*, vol. 12, 2001.

[OCO 98] O'CONNOR D.O., BURKE D.W., BRAGDON C.R., RAMAMURTI B.S., HARRIS W.H., "Dynamic frictional comparison of different femoral head sizes and different lubricants", *Journal of Biomechanics*, 31, 1998

[OON 09] OONISHI H., UENO M., KIM S.C. *et al.,* "Ceramic versus cobalt-chrome femoral components; wear of polyethylene insert in total knee prosthesis", *J. Arthroplasty*, vol. 24, pp. 374-382, 2009.

[PAU 66] PAUL J.P., "Forces transmitted by joints in the human body", *Proceedings of Institution of Mechanical Engineers*, vol. 181 no. 37, pp. 8-15, 1966.

[RIE 05] RIEKER C.B., SCHON R., KONRAD R., LIEBENTRITT G., GNEPF P., SHEN M., ROBERTS P., GRIGORIS P., "Influence of the clearance on in-vitro tribology of large diameter metal-on-metal articulations pertaining to resurfacing hip implants", *Orthopedic Clinics of North America*, 36(2):135, 2005.

[SAI 09] SAIKKO V., "Friction measurement in the biaxial rocking motion hip joint simulator", *Journal of Tribology*, vol. 131, 2009

[SCH 00a] SCHOLES S.C., UNSWORTH A., "Comparison of friction and lubrication of different hip prostheses", *Proceedings of the Institution of Mechanical Engineers, Part H, Journal of Engineering in Medicine*, vol. 214, no. H1, pp. 49-57, 2000.

[SCH 00b] SCHOLES S.C., UNSWORTH A., GOLDSMITH A.A., "A frictional study of total hip joint replacements", *Physics in Medicine and Biology*, vol. 45, no. 12, pp. 3721-3735, 2000.

[SCH 01] SCHOLES S.C., GREEN S.M., UNSWORTH A., "The wear of metal-on-metal total hip prostheses measured in a hip simulator", *Proceedings of the Institution of Mechanical Engineers, Part H, Journal of Engineering in Medicine*, vol. 215, no. 6, pp. 523-530, 2001.

[SCH 06] SCHOLES S.C., UNSWORTH A., "The effect of proteins on the friction and lubrication of artificial joints", *Proceedings of the Institute of Mechanical Engineers, Part H, Journal of Engineering in Medicine*, vol. 220, pp. 687-693, 2006.

[SMI 01] SMITH S., DOWSON D., GOLDSMITH A.A.J., VALIZADEH R., COLLIGON J.S., "Direct evidence of lubrication in ceramic-on-ceramic total hip replacements", *Proceedings of the Institution of Mechanical Engineers*, vol. 215 Part C, pp. 265-268, 2001.

[STE 03] STEWART T.D., TIPPER J.L., INSLEY G., STREICHER R.M., INGHAM E., FISHER J., "Long-term wear of ceramic matrix composite materials for hip prostheses under severe swing phase microseparation", *JBMR*, Part B, vol. 66, 2001.

[UDO 03] UDOFIA I.J., Tribology of metal on metal hip resurfacing prostheses, PhD Thesis, University of Bradford, 2003.

[UTZ 09] UTZSCHNEIDER S., HARRASSER N., SCHROEDER C., MAZOOCHIAN F., JANSSON V., "Wear of contemporary total knee replacements – A knee simulator study of six current designs", *Clinical Biomechanics*, vol. 24, pp. 583-588, 2009.

[WIL 96] WILLERT H.G., SEMLITSCH M., "Tissue reactions to plastic and metallic wear products of joint endoprostheses", *Clinical Orthopaedics & Related Research*, vol. 333, pp. 4-14, 1996.

[WIL 03] WILLIAMS S., BUTTERFIELD M., STEWART T., INGHAM E., STONE M., FISHER J., "Wear and deformation of ceramic-on-polyethylene total hip replacements with joint laxity and swing phase microseparation", *Proceedings of the Institute of Mechanical Engineers, Part H, Journal of Engineering in Medicine*, vol. 217, 2003.

[WIL 07] WILLIAMS S., SCHEPERS A., ISAAC G., HARDAKER C., INGHAM E., JAGT D.V.D., BRECKON A., FISHER J., "The 2007 Otto Aufranc award: ceramic-on-metal hip arthroplasties: a comparative in vitro and in vivo study", *CORR*, 2007.

[WIL 08] WILLIAMS S., LESLIE I., ISAAC G., JIN Z., INGHAM E., FISHER J., "Tribology and wear of metal-on-metal hip prostheses: influence of cup angle and head position", *J. Bone Joint Surg. Am.*, vol. 90, pp. 111-117, 2008.

[WIM 03] WIMMER M.A., SPRECHER C., HAUERT R., TAGER G., FISCHER A., "Tribochemical reaction on metal-on-metal hip joint bearings: A comparison between in-vitro and in-vivo results", *Wear*, vol. 255 no. 7-12, pp. 1007-1014, 2003.

[YOU 00] YOUNG S.K., KELLER T.S., GREER K.W., GORHAN M.C., "Wear testing of UHMWPE tibial components: influence of oxidation", *J. Tribology*, vol. 122, pp. 323-331, 2000.

Chapter 3

Influence of Temperature on Creep and Deformation in UHMWPE under Tribological Loading in Artificial Joints

This chapter presents an overview of the influence of temperature on the friction and creep characteristics of the ultra-high molecular weight polyethylene used in artificial joints. Additionally, the key findings of *in vitro* tribological studies on the deformation mechanisms of polyethylene are presented and a new model of deformation is introduced. The significance of the results is discussed in relation to artificial joints and their testing.

3.1. Temperature in artificial joints

3.1.1. *Artificial knee joints*

Ultra high molecular weight polyethylene (UHMWPE) combined with cobalt-chromium-molybdenum (CoCrMo)-

---

Chapter written by Mathias Christian GALETZ and Uwe GLATZEL.

alloys has been used as a bearing couple in artificial knee joints for 40 years because of its good balance of properties, such as impact behavior, wear resistance and biocompatibility [KUR 04]. Improved design, surgical technique and material processing have increased the average knee prosthesis survival rate to about 95% after 10 years *in vivo* [SWE 08]. Due to this successful history about 136,500 new total endoprostheses of the knee joint were implanted in Germany in 2007, but unfortunately about 9,500 had to be revised [BQS 07]. Although knee arthroplasty has a remarkable clinical track record, problems with wear and fatigue continue to limit the longevity of knee replacements [KUR 04]. Figure 3.1 shows a modern artificial knee joint. Characteristically for the knee joint there are two condyles, condylus medialis and condylus lateralis. These condyles are substituted by a metallic component in artificial knee joints (Figure 3.1). In-between the femur and the tibia two menisci are found in the natural knee. They are missing in the artificial knee and the metallic condyles slide directly on a plastic inlay, which substitutes the tibia surface.

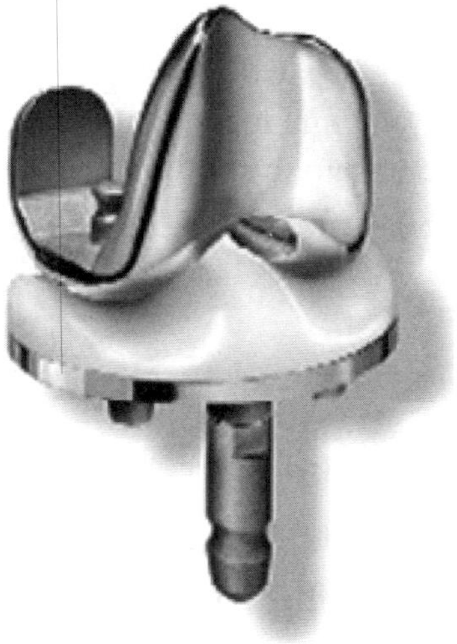

Figure 3.1. *Artificial knee joint (Aequos GmbH)*

The polymer of choice for all tibia inlays is UHMWPE. The tibia inlay is usually supported by a metallic tibia tray. Since the introduction of artificial knee joints in the early sixties all implants rely on this soft/hard bearing concept, in contrast to the arthroplasty of hip joints for which hard/hard material combinations made from metals or ceramics are used.

3.1.2. *Why does temperature affect the performance of artificial joints?*

It is often assumed that artificial joints operate at body temperature. Amazingly, *in vivo* measurements have shown a significant temperature rise in artificial joints. In the joint capsule of artificial knees, a temperature increase of up to $\Delta T = 9$ K after 40 min of walking was measured in comparison to healthy knees [PRI 05]. In artificial hip joints temperatures up to 43°C within the ball were measured *in vivo* [BER 01], in hip simulator tests even more, for example 60°C when running in 45 ml at 1.5 Hz [DAV 87, LIA 03].

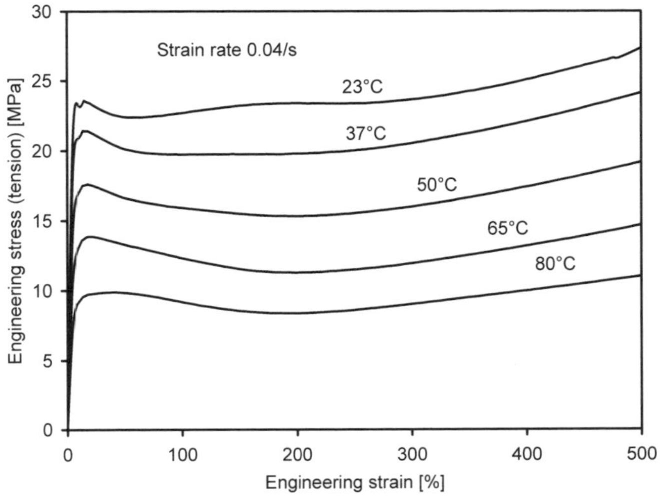

Figure 3.2. *Engineering stress-strain diagram under tension as a function of testing temperature (strain rate: 50 mm / min)*

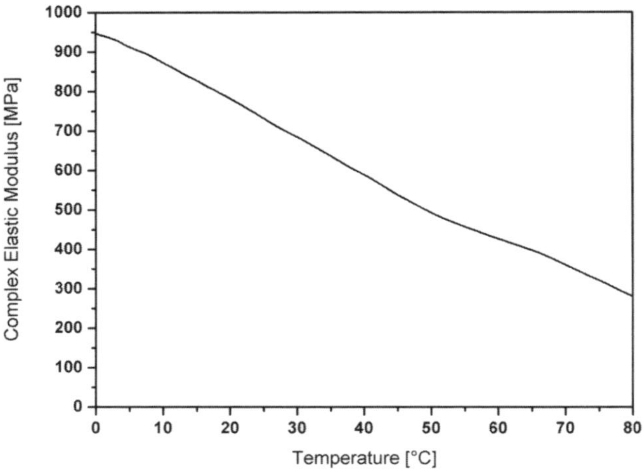

Figure 3.3. *Temperature dependence of the storage and loss modulus of UHWMPE*

Unfortunately, the mechanical properties of UHMWPE are highly temperature dependent. With increasing temperature the strength of the UHMWPE decreases significantly, as can be seen from a temperature-dependent engineering tensile stress–strain diagram [GAL 09b] (Figure 3.2).

In addition to strength, stiffness is also strongly affected by temperature changes. The elastic modulus decreases with temperature because the amorphous phase of the polymer becomes more mobile, even at the small strains investigated by the dynamic mechanical analysis (DMA), see Figure 3.3 [GAL 07].

Generally, the friction force of a visco-elastic polymer such as UHMWPE against a hard rigid body is the sum of two contributions, described as the adhesion force and the hysteretic or deformation model of friction [UET 85, ACH 01]. In the case of adhesion, the applied frictional work is considered to be dissipated in the area of intimate solid-solid contact and corresponds to a high local energy concentration. The hysteretic component prevails when rolling is applied. The surface becomes deformed and frictional work is dissipated due to the internal damping during cyclic deformation [GRE 60]. This friction mode occurs within a larger volume of material and is not restricted to the surface [BRI 81]. It strongly depends on the mechanical properties. Wear is, of course, also linked to the mechanical properties. This clearly shows the importance of investigating the actual temperature in the tribological contact of artificial joints *in vivo*.

3.1.3. *Mathematical approaches to estimate the contact temperature during friction*

Since operating temperatures of artificial joints *in vivo* can only be measured at enormous experimental costs [BER 01] or using medically dubious techniques [PRI 05], simulation and calculation are important tools for estimating the temperature *in vivo*. A first approach to calculating the temperature rise and heat flow in artificial knee joints was made by Sawyer *et al.* [SAW 03], using dynamic contact and thermal analyses.

For a two body system the rate of total energy generated in the contact under friction can be determined by the relative velocity of the counter bodies' v_{rel} and the frictional force, which equals the product of normal load F_n and friction coefficient μ [SAW 03]. Tribologists usually agree that almost all generated frictional power $\dot{Q}$ is transformed into heat [UET 85]:

$$\dot{Q}_{total} = \dot{Q}_{wheel} + \dot{Q}_{inlay} = \mu F_{n} v_{rel} \qquad [3.1]$$

In order to predict surface temperatures, the fraction of the heat flowing into the polymer must be determined. In this case a heat participation factor ε is used, which is defined as:

$$\varepsilon = \dot{Q}_{inlay} / \dot{Q}_{total} \quad [3.2]$$

Under stationary conditions the heat flow into two semi-infinite bodies in contact is partitioned between them according to their thermal conductivities [CAR 59]:

$$\varepsilon_{inlay} = \frac{\lambda_{UHMWPE}}{\lambda_{UHMWPE} + \lambda_{CoCrMo}} \quad [3.3]$$

In their analysis of the temperature in knee joints Sawyer *et al.* [SAW 03] extensively discussed the heat partitioning. They finally used a stationary partitioning. In the following it will be shown that this amount of energy dissipated and therefore the calculated temperature within the polyethylene component during articulation against cobalt-chromium-molybdenum (CoCrMo) is too low. The fraction of friction energy entering the polyethylene is higher because of the movement of the counter bodies.

Most surface temperature analysis [SAW 03, SAM 04, IMA 04] follows the general theory of moving heat sources introduced by Blok [BLO 37] in 1937 and further developed by Jaeger and Carslow [CAR 59]. This theory describes the heat development between two semi-infinite moving bodies. They describe a high dependence of the heat partitioning on the dimensionless Peclet number Pe that may be interpreted as the ratio of the speed of the surface to the rate of diffusion of heat into the solid [BHU 01]. This means that at large Pe, heat can only diffuse a short distance into the solid in the time taken for the surface to move through the heated zone [BHU 01]:

$$Pe_{inlay/wheel} = \frac{v_{inlay/wheel}\, b}{2\chi} \quad [3.4]$$

where v_{inlay} or v_{wheel} is the translative velocity [m/s] of the respective counter body, χ [m²/s] is the thermal diffusivity, i.e. $\chi = \lambda/(c_p\rho)$ with λ [W/mK] being the thermal conductivity, ρ [kg/m³] the density, c_p [J/kgK] the specific heat and b [m] a contact dimension [m]. For the wheel-on-flat contact, the contact width b is the half width of the rectangle for linear contacts. It is calculated according to the Hertzian equation:

$$b = \sqrt{F_n r_{wheel} / l\pi E^*} \qquad [3.5]$$

F_n is the normal load, r_{wheel} the radius of the wheel, l the contact length and E^* the contact modulus defined by:

$$\frac{1}{E^*} = \frac{1-\upsilon_{CoCr}^2}{E_{CoCr}} + \frac{1-\upsilon_{UHMWPE}^2}{E_{UHMWPE}} \qquad [3.6]$$

The required material properties for the common material combination used in artificial knee joints are given in Table 3.1.

	UHMWPE	CoCrMo
Thermal conductivity [W/mK]	0.4 [IMA 04]	12.1 [IMA 04]
Thermal diffusivity [m²/s]	2.4e-7 [IMA 04]	3.4e-6 [IMA 04]
Young's modulus at 37°C [MPa]	800 [GAL 07]	200000 [IMA 04]
Poisson ratio	0.46 [KUR 99]	0.3 [IMA 04]

Table 3.1. *Mechanical and thermal properties of UHMWPE and CoCrMo*

It can easily be seen that the contact width itself is a function of temperature, due to the high temperature dependence of the modulus of UHMWPE, shown in Figure 3.3 [GAL 07]. For example, for the wheel-on-flat testing setup shown in Figure 3.4 at a load of 1.3 kN, the contact width changes significantly as a function of temperature (Figure 3.5), based on the change of the modulus of polyethylene shown in Figure 3.3.

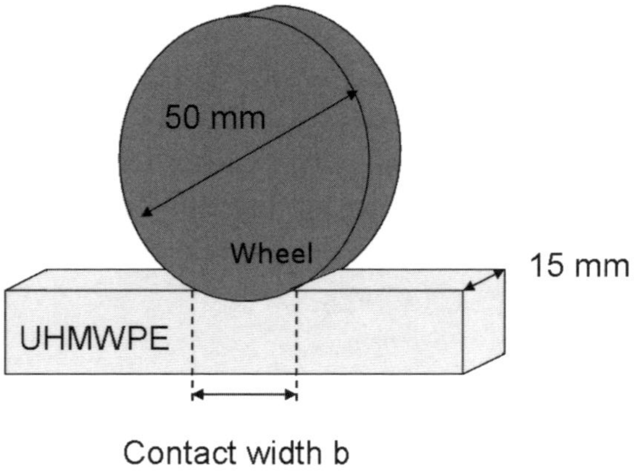

Figure 3.4. *Contact width for a wheel-on-flat configuration*

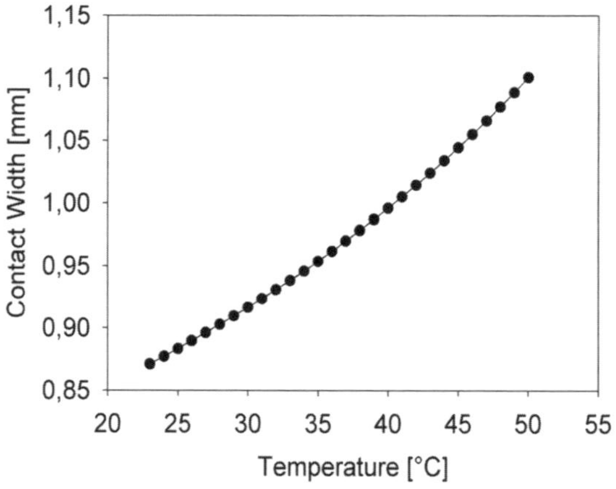

Figure 3.5. *Calculated contact width as a function of temperature*

However, assuming that the maximum temperatures of both surfaces are the same, the proportion of thermal energy into UHMWPE can be calculated by equation [3.6] for Peclet numbers >1 [BHU 01]:

$$\varepsilon = \frac{1}{1 + \frac{\lambda_{CoCr}\sqrt{Pe_{CoCr}}}{\lambda_{UHMWPE}\sqrt{Pe_{UHMWPE}}}} \qquad [3.7]$$

The mean temperature rise at the surface for both relative velocities can then be estimated according to Jaeger and Carslow [BHU 01]:

$$\Delta T = \frac{4\varepsilon \dot{Q}_{total} b}{3A\lambda_{UHMWPE}\sqrt{\pi\ Pe_{UHMWPE}}} \qquad [3.8]$$

The most significant restriction of this theory of moving heat sources is that it does not cover reciprocating sliding and thus cannot be assigned to cyclic friction of polymers [SAM 04, IMA 04] in reciprocal tribological testing devices or knee simulators. Instead, this model is only valid for unidirectional sliding [SAM 04, IMA 04] and gives the temperature rise in the contact during one single cycle.

3.1.4. *Temperature rise during cyclic tribological sliding*

The temperature rise during a constant reciprocating sliding on a wheel-on-flat simulator set-up with three degrees of freedom, described in the literature [SCH 05, GAL 09a,b] was investigated and is shown schematically in Figure 3.6.

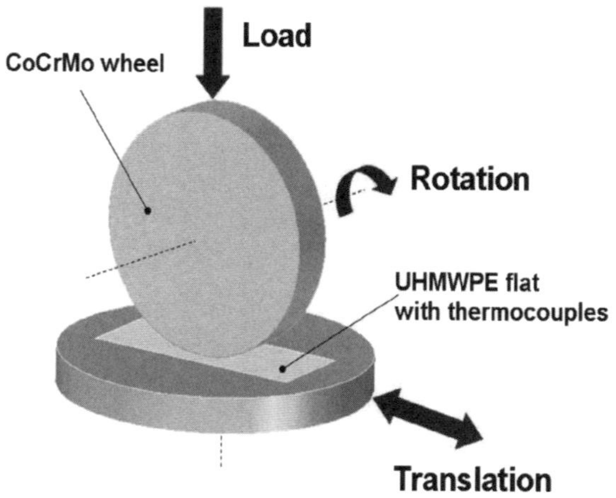

Figure 3.6. *Principle of the wheel-on-flat tester for the temperature rise in UHMWPE inlays (reprinted from [GAL 09a] © 2009, with permission from Elsevier)*

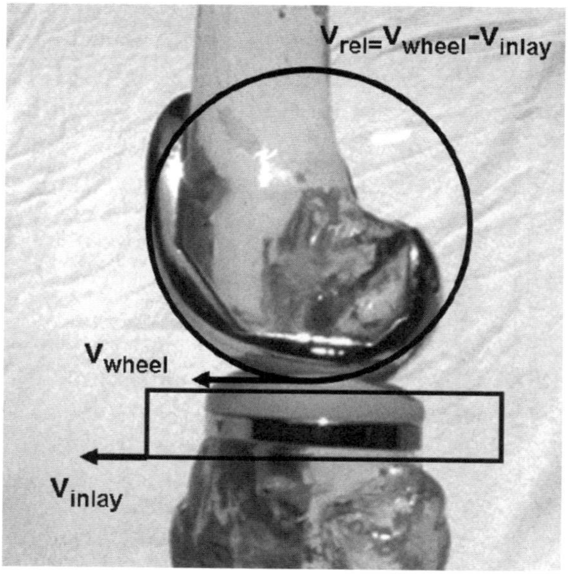

Figure 3.7. *Relative velocity of a knee joint and its relation to a wheel-on-flat system (reprinted from [GAL 09a] © 2009, with permission from Elsevier)*

With this setup a wide range of tribological loading conditions can be investigated, which can occur in knee joints. In particular, the relative velocity between both counterbodies can be changed. This is defined in Figure 3.7 for an artificial knee joint or wheel-on-flat. In the wheel-on-flat tests a CoCrMo wheel (cast CoCr28Mo6) with a diameter of 100 mm and a thickness of 20 mm articulated against an UHMWPE flat (Chirulen 1020).

In the tests described in the following a sinusoidal friction power profile was applied on the inlay. The UHMWPE inlay was moved with an amplitude of translation motion of 10 mm at a maximum velocity of 26 mm/s. By defining different angular velocities of the wheel, the relative velocity in the contact varied between 0 and 182 mm/s. The normal load had a sinusoidal form and was set to 1.3 kN to mimic one condyle. The temperature change inside the polyethylene component was recorded with embedded thermocouples. These measurements allow a determination of the temperature rise. Additionally the dependence of the coefficient of friction on temperature was determined. The exact testing parameters can be found elsewhere [GAL 09a, GAL 09b].

A significant temperature rise is found in UHMWPE under this sinusoidal loading condition when sliding against CoCrMo. In dry sliding a temperature increase of up to $\Delta T = 45$ K was observed 4 mm beneath the surface, depending on the relative velocity (Figure 3.8).

To show the large difference between reciprocal sliding and one translation of the wheel the temperature rise at the immediate contact of the surface was calculated according to equation [3.7]. For the smallest relative velocity, shown in Figure 3.8, it is only $\Delta T_{26mm/s}$= 0.12 K per cycle right at the surface, compared to the 5°C, reached 4 mm under the contact after 3,600 cycles. The value per cycle is much smaller than expected and shows the huge difference between one cycle and the accumulated temperature rise during reciprocating sliding.

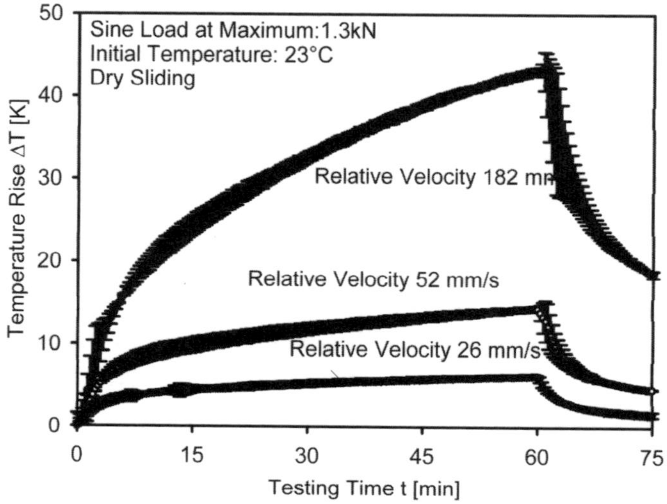

Figure 3.8. *Temperature rise 4 mm beneath the surface throughout the test and 30 min after testing (reprinted from [GAL 09a] © 2009, with permission from Elsevier)*

It was also shown that the temperature rise highly depends on the amount of liquid present. If we changed from dry sliding to about 100 ml serum or water only about 1 K was measured compared to 43 K under dry sliding. Using a volume of 250 ml, none of the two liquids led to a detectable temperature rise inside the polyethylene. This proves that the temperature rise can be suppressed by cooling through the large amount of liquid.

Using 0.2 ml paraffin oil as a lubricant the coefficient of friction is in the same order as with serum or water but a temperature increase of 4 to 5 K is found during sliding (Figure 3.9), that directly reflects the one order of magnitude difference in the coefficient of friction, which is 0.13 in dry sliding and 0.015 when lubricated with paraffin oil for the same relative velocity.

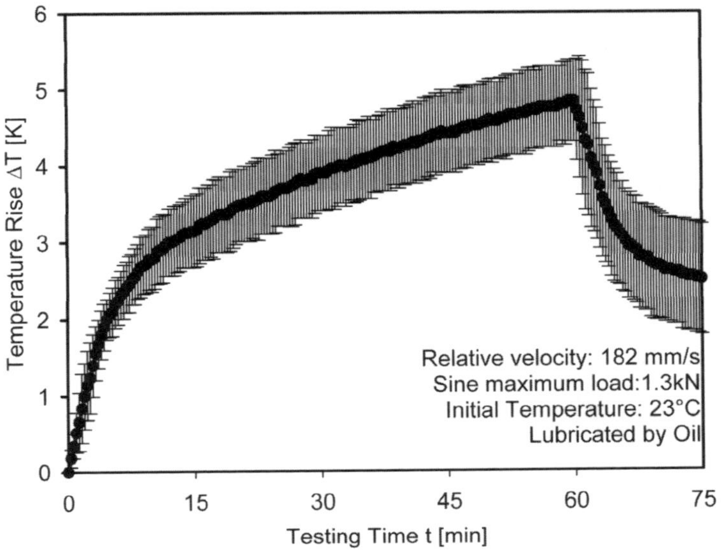

Figure 3.9. *Temperature rise 4 mm under the surface (oil lubrication) (reprinted from [GAL 09a] © 2009, with permission from Elsevier)*

In contrast, rolling, with no relative velocity between the wheel and the flat, led to a temperature increase of only $\Delta T = 0.3$ K under lubricated conditions and even in dry sliding. During rolling without relative velocity adhesive effects play a minor role. This shows that the contribution of heat dissipation due to reciprocal loading is detectable within the bulk, but plays a minor role compared to the heat generated close to the surface.

Therefore, adhesion and hysteretic heating must operate close to the surface and lead to the high temperature differences within the material. Under all investigated loading conditions with sinusoidal load, the temperature rise is linear related to the friction energy given by load, coefficient of friction and relative velocity (Figure 3.10).

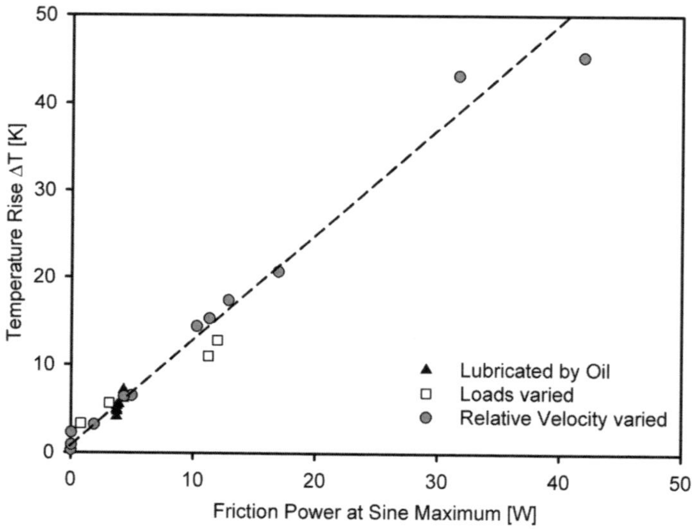

Figure 3.10. *Steady-state temperature rise under sinusoidal loading as a function of the friction energy at* ($\mu F_n v_{rel}$) *maximum load (reprinted from [GAL 09a] © 2009, with permission from Elsevier)*

The coefficient of friction after the running-in was used for the calculation of the friction energy. This coefficient of friction depends on the relative velocity in the tests under dry sliding conditions, but for all different contact speeds it became stable after the first 1,000 cycles (Figure 3.11). This indicates that the conditions close to the surface remain constant afterwards, including the surface temperature. Yet after 15 min the temperature still rises in the bulk. The explanation for the initial increase of the coefficient of friction is that the true contact area increases in the early stage of running-in. This leads to a quick smoothing of its surface, combined with a more uniform distribution of pressure at the contact surface and higher adhesive forces [TAN 77]. Especially for smooth polymers, the interaction between both counter bodies is dominated by adhesion forces [FUR 03]. A very low coefficient of friction at the beginning was found, that might even be the result of the melting of asperities on the UHMWPE [WIE 07]. The observed maximum in the coefficient of friction as a function of relative velocity and load is a typical sign that friction due to deformation also has a large influence [KRA 83].

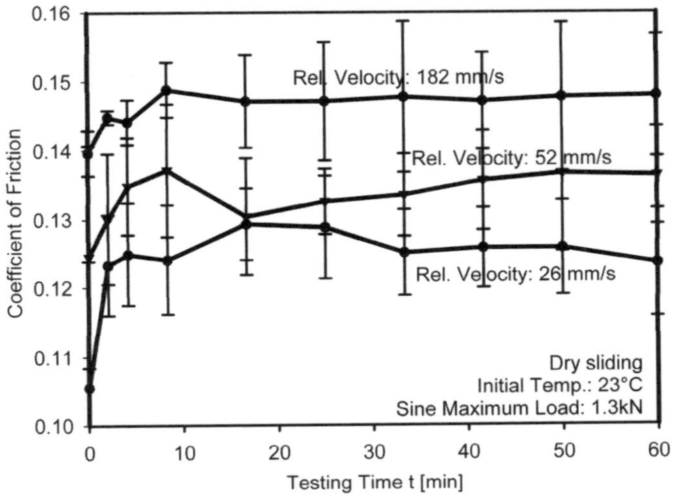

Figure 3.11. *Alteration of the coefficient of friction throughout the test duration of 60 min (reprinted from [GAL 09a] © 2009, with permission from Elsevier)*

By applying a rectangular load instead of a sinusoidal load, the heat fraction dissipated in the polyethylene could be determined, as described elsewhere [GAL 09a]. Only a small variation for different relative velocities between 0.074 and $\varepsilon_2 = 0.065$ for a fixed wheel or one moving with an angular speed of about 100 mm/s was observed. The little dependency on the velocity of the wheel indicates that most of the energy is dissipated within the polymer not directly in the contact zone at the surface. In the first case (v_{rel} = 26 mm/s) the wheel is fixed, therefore the same part of it is always in contact with the polyethylene.

In the second test (v_{rel} = 78 mm/s) 80 mm of the circumference of the wheel came into contact during each cycle, which can increase its cooling potential. On the other hand, with higher translation movement less heat flows into the CoCrMo wheel according to Carslow and Jaeger. While this theory also involves a high dependence of the heat participation on relative velocity, it clearly cannot be directly applied to reciprocating sliding relevant to knee joints. The heat participation remains almost stable not only in the test with rectangular loading, as can be concluded from the test with sinusoidal loading (Figure 3.8). Therefore it is suggested that the fraction of heat entering the polyethylene can be taken to be about 0.07 of the total friction energy and approximately constant in the load range of artificial joints.

This facilitates calculations of the temperature in knee joints, but it is twice the amount (0.035) that was used in a theoretical analysis of Sawyer *et al.*, who calculated a temperature rise of 9°C during continuous activity of gait [SAW 03]. As the temperature increase is proportional to the fraction of heat, the model of Sawyer *et al.* in combination with the present results suggests that the temperature increase is 18°C directly at the surface.

3.2. Temperature influence on creep and fatigue mechanisms of UHMWPE under tribological loading

3.2.1. *Temperature dependence of the yield strength of UHMWPE*

Variations of some degrees above 37°C have a high impact on the performance of UHMWPE inlays, because the polymer is operated close to the softening temperature of about 80°C. One the one hand, when the temperature in the friction contact of artificial joints rises above 37°C, it can lead to accelerated creep and failure [DON 07], compared to *in vitro* tests run in a rather large amount of liquid and therefore a stable temperature of 37°C. On the other hand, if the temperature influence is predictable, testing at elevated temperature can provide a possibility to reduce testing times *in vitro*. Walking activity studies show that patients with artificial joints do about 5,000 steps per day on average, which extrapolates to approximately 1 million cycles per year for each joint of the lower extremity [SCH 99]. *In vitro* creep or stress relaxation can only be determined for an experimentally limited duration. The tests to simulate the lifetime of artificial joints are typically done for some million cycles at 1 Hz and hence are very time consuming and expensive.

Generally, the principle of testing at higher temperatures is used to accelerate the creep behavior of polymers. One example is the internal "rupture test" of polyethylene pipes. By increasing the temperature, an extrapolation of the measured data at higher temperature to many years in service at lower temperatures is possible in compliance with the application of the Arrhenius rate law [BAR 79, BER 07]. The model is based on the assumption that the polymer deformation obeys the general Arrhenius equation for molecular kinetics. The Arrhenius equation is a simple and remarkably accurate formula for the temperature dependence of many reaction rates. Based on that principle an enhanced theory was developed by Eyring in 1936 [EYR 36]. This theory also includes the stress dependence to describe the yielding and creep of polymers as a thermally activated process. Following the formalism developed for creep under constant applied stress, the creep rate is given by the following equation:

$$\dot{\varepsilon} = \dot{\varepsilon}_0 \exp\left(\frac{-\Delta E_a}{kT}\right)\sinh\left(\frac{\sigma V}{kT}\right) \qquad [3.9]$$

ΔEa is the activation energy and σ is the stress (usually tensile or compressive); V is the activation volume, governing the effectiveness of the stress in reducing the activation barrier and ε_0 is a constant pre-exponential factor, related to the fundamental jumping frequency for the flow units to bypass each other and k is the Boltzmann constant. As long as the underlying deformation mechanisms do not change, similar results can be achieved in short time at an increased temperature as in tests at a lower temperature with longer times.

The strain-rate and the pressure dependence of the yield stress [SEQ 94, FOT 78, TRU 84] of polyethylene as well as the creep behavior [WIL 78] have already been studied using the Eyring theory. Polyethylene usually shows two yield points, one at about 0.4% and one at higher strains. The first yield point was exclusively associated with the amorphous fraction [DAS 03], but can be neglected at high stresses, relevant in artificial joints.

The mechanisms relevant for the second yield point and under creep at high loads were ascribed to the crystalline fraction and the generation and propagation of screw dislocations [YOU 73, SHA 76, CRI 89, SEQ 02] with a contribution of the amorphous fraction deformed by shear yielding and/or crazing [FRI 83]. It is hypothesized that within a certain temperature range it is possible to accelerate the creep under tribological loading without changing the underlying deformation mechanisms. As was shown in Figure 3.2 for tensile tests, the yield point of UHMWPE strongly depends on strain temperature. In Figure 3.12 similar results are shown, but for quasi-static compression tests. Additionally, the strain rate dependence is shown in Figure 3.13. From these diagrams the true yield points can be calculated [GAL 10a] and these true yield points per temperature can be plotted as a function of the strain rate in the "Eyring plot", given in Figure 3.14. From this slope and the distance of the lines in the Eyring plot, the activation energy of the process can be calculated using equation [3.9]. The exact calculation procedure is given elsewhere [GAL 10a].

Figure 3.15 shows that the activation energy remains stable within the temperature range investigated, while the activation volume changes at temperatures above 50°C and the deformation under stress increases. This means the deformation in compression can be described by one thermally activated Eyring process and this is a very good indication for a stable deformation mechanism between 23 and 50°C.

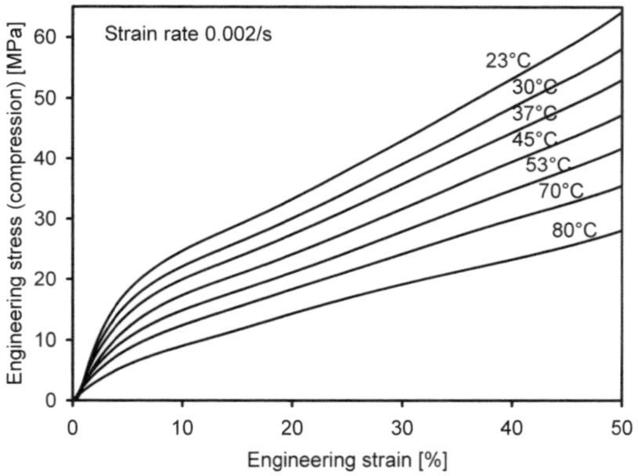

Figure 3.12. *Engineering stress-strain diagram under compression as a function of testing temperature (strain rate: 5 mm / min) (reprinted from [GAL 10a] © 2009, with permission from Elsevier)*

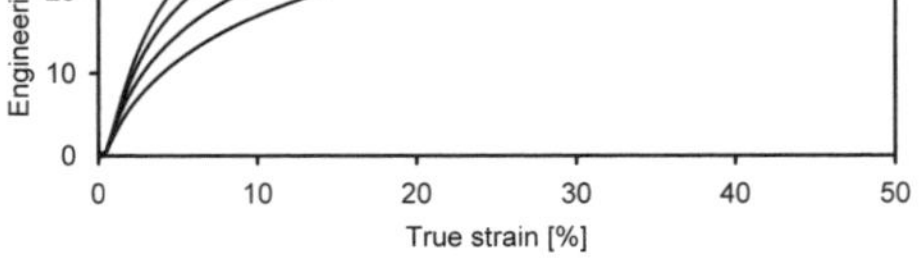

Figure 3.13. *Engineering stress-strain diagram under compression as a function of testing strain rate at 23°C (reprinted from [GAL 10a] © 2009, with permission from Elsevier)*

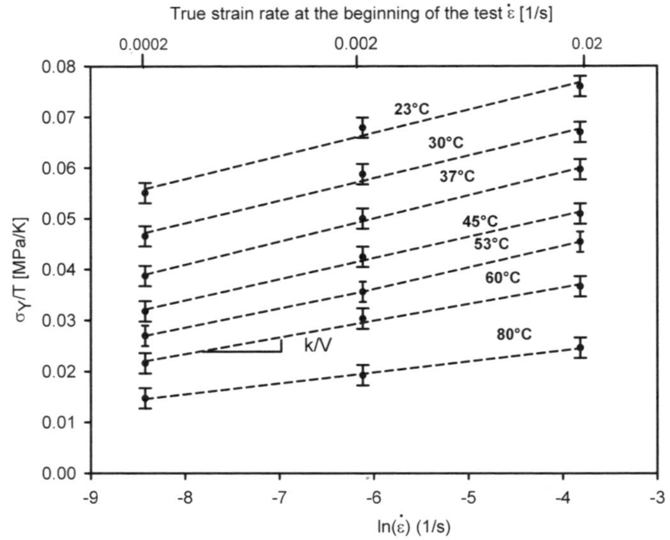

Figure 3.14. *Temperature and strain rate dependent yielding of UHMWPE under compression (reprinted from [GAL 10a] © 2009, with permission from Elsevier)*

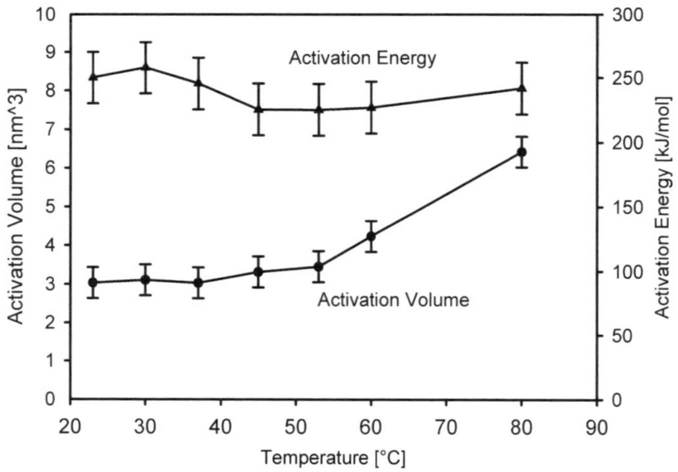

Figure 3.15. *Temperature dependence of the activation energy and activation volume of the yield point under compression (reprinted from [GAL 10a] © 2009, with permission from Elsevier)*

The underlying change in the mechanism above 50°C is regarded as the same that is known as the mechanical α-relaxation of linear high density polyethylene, and assigned to more mobile chains within the crystalline phase due to higher free volume [ZHO 98].

3.2.2. *Temperature dependence of the creep strength of UHMWPE*

These findings on the yield stress are in good agreement with results from creep tests under compression, where samples were loaded at a constant load of 10 and 20 MPa for 300,000 s.

Figure 3.16 shows that in-between 23 and 45°C the permanent strain after the test depends linearly on the temperature while above 50°C extensive additional flow is observed. This effect is more pronounced at 20 MPa. It is in good agreement with the changed activation volume above 50°C that is variable for the effectiveness of the stress. This suggests that the observed creep is linked to the same Eyring flow process observed in the static compression tests.

From these findings it is concluded that when mainly compression loads act, the deformation speed can be accelerated by increasing the temperature up to 50°C. Above this temperature the mechanism changes and wrong conclusions might be drawn for the deformation. It must also be mentioned that this approach is limited to compression stresses.

Under tensile stresses the deformation is more complex. At lower temperatures whitening of the polymer during deformation is observed, which is explained by crazing. Between 23°C and 50°C the heaviest crazing is observed for 23°C.

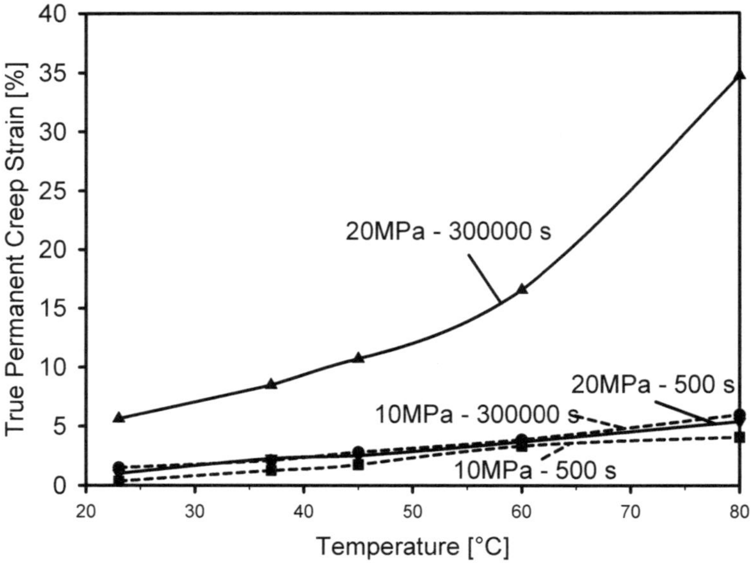

Figure 3.16. *Permanent strain after 500 s and 300,000 s (reprinted from [GAL 10a] © 2009, with permission from Elsevier)*

This crazing is a form of deformation, which can only be partially recovered, due to the high amount of micro voids. To explain this behavior, it was suggested that under tensile load two Eyring processes are necessary for sufficiently describing the temperature dependence of the yield point of high density polyethylene [SEQ 94].

At higher temperatures, above 40°C, the development of crazes becomes irrelevant and the deformation mechanism is shear yielding of the amorphous phase in combination with dislocation motion in the crystalline fraction, similar to that under compression load. Therefore at temperatures relevant to artificial knees, the approach of accelerated deformation at higher temperatures up to 50°C should be applicable, even for complex loads as under tribological loading.

3.2.3. *Temperature-dependent deformation under tribological loads*

To investigate the influence of temperature on creep under tribological loads, tests were run in distilled water at 1 Hz using a three station-wheel-on-flat testing device. A single axis twin peak type loading curve [PAU 76] was applied on the CoCrMo-wheels through the vertical axis of the simulator (Figure 3.17).

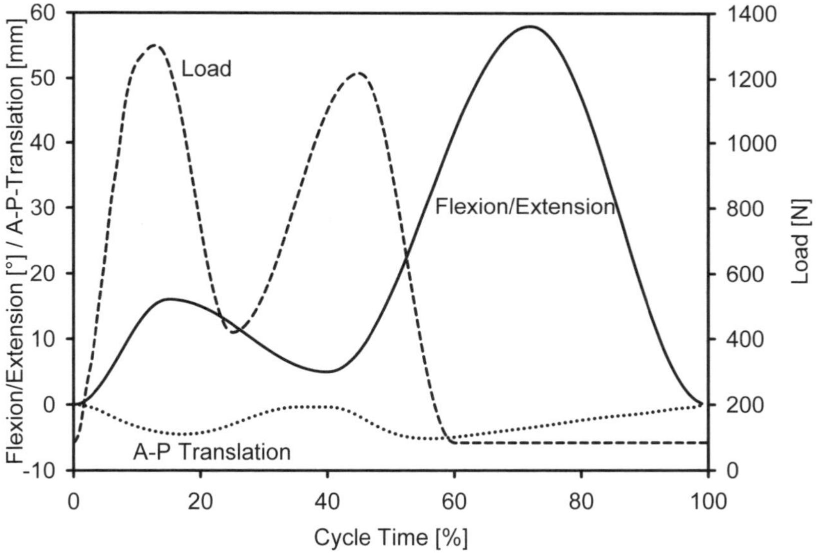

Figure 3.17. *Loading parameters of the testing device shown in Figure 3.18*

The flexion-extension motion of the knee was realized according to ISO-14323-3 by rotating the wheels, while the anterior-posterior-translation was applied by moving the UHMWPE-flat (Figure 3.18). The rotational movement about a vertical axis that is present in the natural knee was omitted. As shown before, this significantly reduces the wear [WAN 96, BRA 96, SAI 01]. The focus of the tests lies on the creep behavior and deformation mechanisms. Low wear is beneficial because extensive wear complicates the determination of the deformation processes close to the surface.

Figure 3.18. *Self-made tribological testing device*

This tribological test proved that deformation can clearly be accelerated by testing at higher temperatures. In Figures 3.19-3.21 the profile of the wear scar along the translation path is shown for 23°C, 37°C and 45°C after 100,000, 1,000,000 and 3,000,000 cycles. Figure 3.22 shows that the maximum depth of the wear scar along the loading path, which gives the magnitude of the observed creep under frictional loading is a function of temperature. The same permanent deformation that occurs after 3 million cycles testing at 23°C was found after about 900,000 cycles at 37°C and even after about 500,000 cycles at 45°C. The same deformation was achieved six times faster at 45°C than at 23°C.

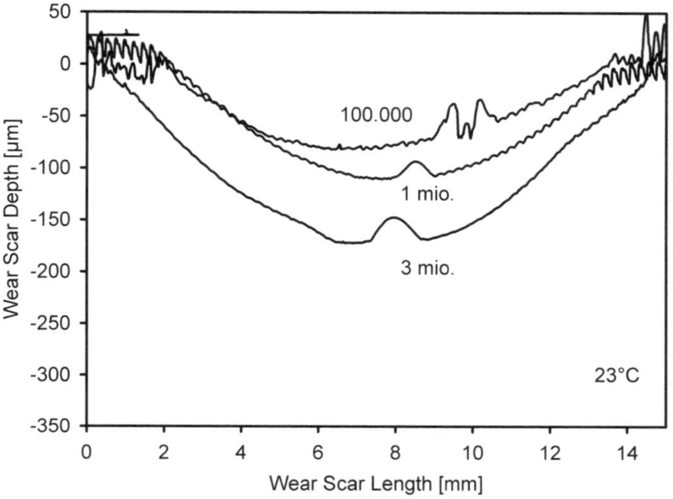

Figure 3.19. *Typical wear profile after 100,000, 1 mio. and 3 mio. cycles, when tested at 23°C (reprinted from [GAL 10a] © 2009, with permission from Elsevier)*

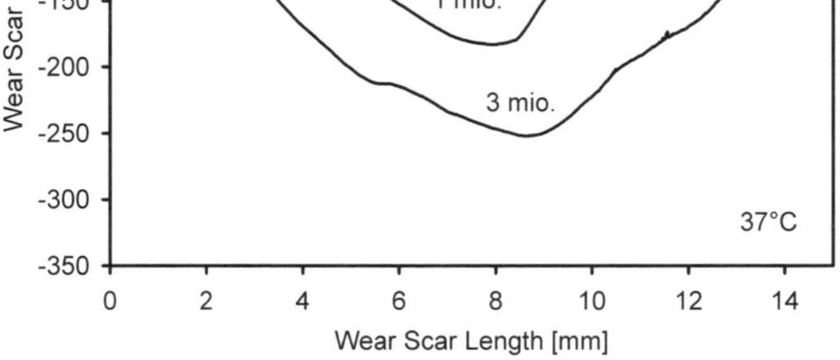

Figure 3.20. *Typical wear profile after 100,000, 1 mio. and 3 mio. cycles, when tested at 37°C (Reprinted from [GAL 10a] © 2009, with permission from Elsevier)*

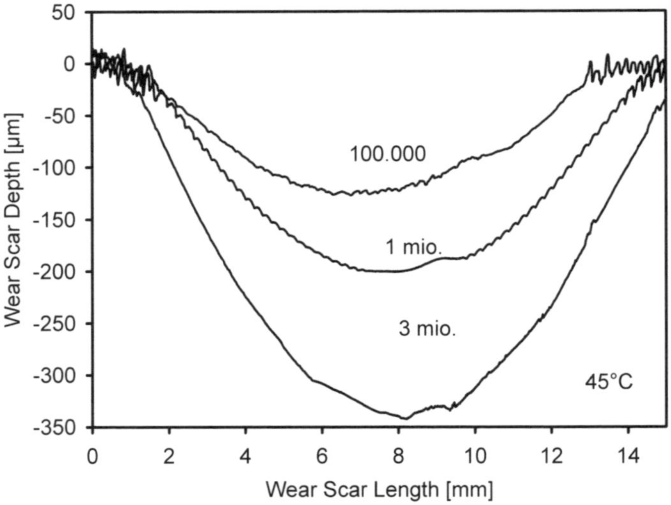

Figure 3.21. *Typical wear profile after 100,000, 1 mio. and 3 mio. cycles, when tested at 45°C (reprinted from [GAL 10a] © 2009, with permission from Elsevier)*

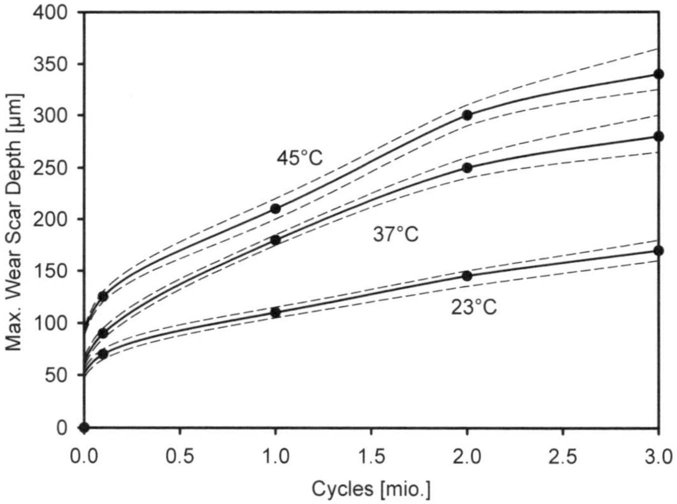

Figure 3.22. *Maximum wear depth of the profile (reprinted from [GAL 10a] © 2009, with permission from Elsevier)*

In the reported tests the temperature was stable within the specimen, due to static testing or by controlling it with a large amount of liquid in the tribological test. Nevertheless higher temperatures are found in artificial joints during walking as shown in section 3.1.1. The present tests show that these elevated temperatures bear a high risk of accelerated and excessive creep *in vivo* especially at intense continuous activities.

As far as wear is concerned in all three tests the weight change was less than 0.0001% and no significant difference between the three temperatures was found. Nevertheless, by investigating the wear track, the deformation and wear behavior of UHMWPE on the molecular scale can be determined, as shown in Chapter 4.

3.2.4. *Wear and deformation mechanisms of ultra-high molecular weight polyethylene*

On the influence of temperature on wear, conflicting reports can be found in the literature. According to Uetz [UET 85] thermal softening induces surface plastification and can lead to high wear rates above a certain point. In contrast a slight temperature increase to 40-42°C was reported not to have a significant effect on the wear rate [EDI 99], and even a decrease of wear with temperature has been observed [LIA 03, LU 97]. This difference is ascribed to the many different types of wear mechanisms which can occur, depending on the exact testing parameters.

Retrieved knee joints show a large variety of wear manifestations. An overview of the numerous literature sources is given by [MCK 07]. From this survey it follows that any visible or microscopic changes in the appearance of the polyethylene surface should be defined as wear damage. Numerous investigators have noted that polyethylene components may exhibit substantial surface damage even though very small weight loss is observed. Therefore, usually the description of the type of wear that occurs on retrieved polyethylene components is limited to the surface appearance.

Phenomenons such as polishing, burnishing, scratches, plastic deformation, cracks, pits or delamination are often described. All these surface changes are more or less believed to be caused by one or more of the classic wear mechanisms generally described for polymers [LAN 79]: abrasion, adhesion and fatigue.

Abrasion is a cutting process of the polymer chains and occurs when hard particles or asperities plough through the polymer [DOW 87, TUR 05].

Adhesion is also effective right at the surface and can even lead to the development of a transfer layer on the metallic counterpart due to high affinity of metals to carbon and hydrogen [GAL 10b,c]. This is typical of polymers such as polyethylene with a smooth molecular profile lacking side groups and kinks in the chain [POO 72, BRI 81].

The formation of the transfer layer depends on the environment and is more severe when tested in water, compared with serum, but it has even been found on retrieved implants [WAL 96]. While abrasion and adhesion take effect close to the surface, in the depths creep mainly occurs. UHMWPE tibial inserts are subjected to periodical physiological loading during walking at peak stresses far greater than the yield strength [BAR 95]. The relevant fatigue failure mechanisms are associated with yielding and accumulated plastic flow processes of UHMWPE [KUR 98, WAN 95].

Implant retrieval analysis suggests that patient weight, activity level, and time of implantation are associated with the severity of component damage [HOO 83]. Wear as a result of creep of fatigue leads to changes in the form and dimensions of the tibia inlays as well as molecular orientations without any loss of mass. This in turn influences the kinematics of the knee joint and therefore the tribological behavior [WIE 07]. It is known that close to the surface a highly oriented layer often occurs in UHMWPE during sliding [MCK 81, CHA 98, MAR 91, CHO 04]. Nevertheless, this cannot explain the macroscopic deformations and is only one aspect of molecular rearrangement.

3.3. Deformation behavior of polyethylene on the molecular scale

3.3.1. *Deformation mechanisms in polyethylene*

No comprehensive picture of the deformation and wear mechanisms on the molecular scale in artificial knee joints exists, in spite of the numerous publications dealing with it.

One reason might be that under stress, several deformation mechanisms occur in polyethylene. These mechanisms have mainly been investigated on high density polyethylene (HDPE) [LIN 94, OLE 03]. They can be transferred to UHMWPE with the limitation of the influence of the spherulitic structure of HDPE that does not exist in UHMWPE.

Basically, the interlamellar mechanisms in the amorphous phase and the intralamellar mechanisms in the crystalline fraction have to be distinguished. This structure can be observed with transmission electron microscopy (TEM) and scanning electron microscopy (SEM), as shown in Figures 3.23 and 3.24.

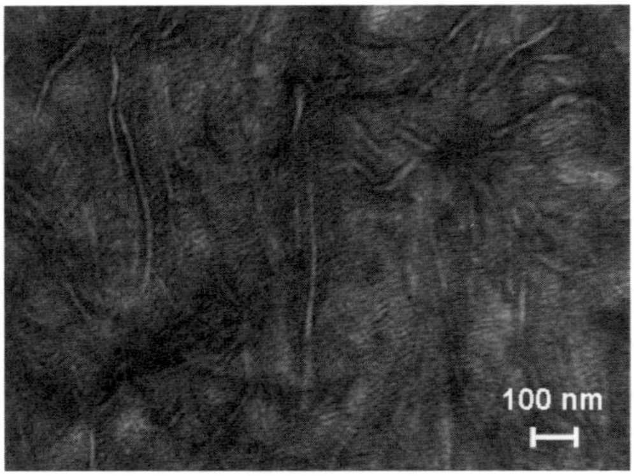

Figure 3.23. *Lamellae structure of UHMWPE (TEM)*

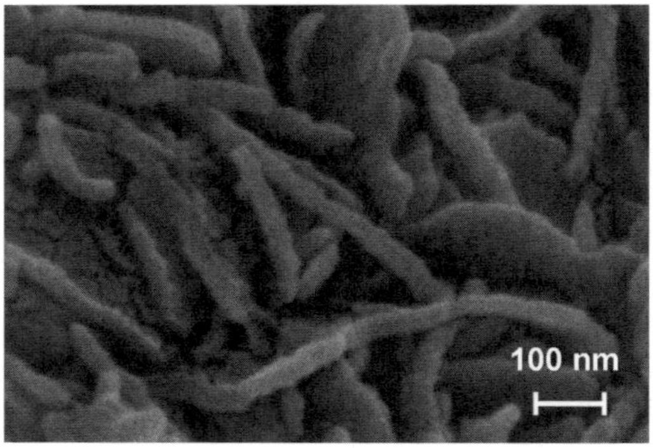

Figure 3.24. *Lamellae structure of UHMWPE (SEM)*

In the beginning, the deformation under stress is restricted to the amorphous phase, because the glass transition temperature of the amorphous phase of polyethylene is well below room temperature (about -160°C [KUR 04]), resulting in a modulus and strength much lower than in the crystalline domains. The activation energy for the deformation of the amorphous phase is just 2-10% of the activation energy needed to induce chain-dislocation-slip within the crystals [PET 66]. It was shown that the deformation of the amorphous phase is reversible up to a strain of 0.4%, mainly due to the entropy-elastic behavior of the molecules [YOU 73, POP 75, PAT 78, BAR 92, BAR 95]. The main mechanisms in the amorphous regions between the crystal lamellae are interlamellar shearing, interlamellar stretching and induced lamella stack rotation [BOW 94].

At higher strains, additional mechanisms of deformation occur in the amorphous phase. One mechanism that only occurs under tension is the development of crazes. Small micro voids can be found, especially at low temperatures or high deformation rates. These micro voids are bridged with molecules, but if the applied tensile load is sufficient, these bridges elongate, break and lead to cracks [FRI 83].

Further mechanisms at high strains all occur within the crystalline part. Within the crystalline lamellae the molecules are in a highly ordered state. In undeformed polyethylene these crystals have an orthorhombic structure; in Figure 3.25 with the lattice parameters a = 740 pm, b = 493 pm and c = 254 pm [LIN 94].

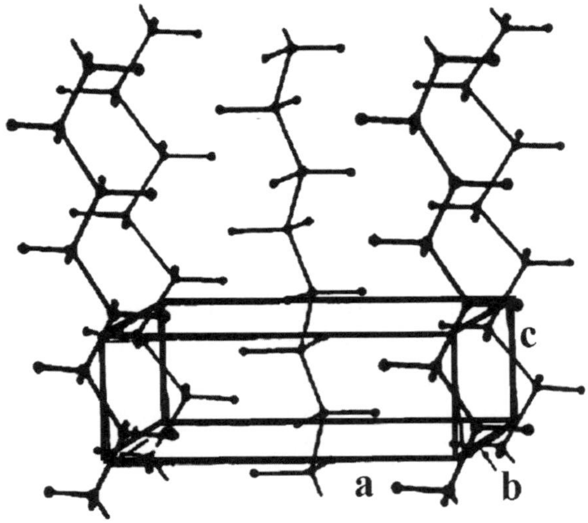

Figure 3.25. *Orthorhombic polyethylene crystal*

The zigzag carbon chains of the molecules are parallel to the crystal c axis, which is perpendicular to the axis of the lamellae. As a result, the long chain molecules must be folded back and forth many times on a plane perpendicular to the axis of the lamellae [LIN 94].

At higher strain, the deformation mechanisms in the crystals always act in combination with the shear of the amorphous part [OLE 03]. In the crystals, lattice shearing from orthorhombic to monoclinic configuration is found. Furthermore mechanical twinning and dislocation motion takes place [POP 75].

At high strains the critical resolved shear stress of dislocation movement is exceeded and slip is induced. Dislocations in polyethylene preferentially emerge around chain ends [PRE 66], but can also occur in continuous chains as was shown by [POP 75].

The long-chain nature of polymer molecules requires that the most preferred slip plane in polymer crystals contains the molecular chain. Under this condition, the polyethylene chains can remain unbroken through very large deformations [KAU 72] and lead to a deformation induced orientation of the crystal structure in polyethylene. The described mechanisms in the literature were mainly investigated under the action of bulk stresses.

Which mechanisms occur and where they appear in the samples of the temperature-dependent wheel-on-flat tests were investigated in the previous chapter. Different zones at certain depths of the samples where investigated, using different methods such as Raman spectroscopy, scanning microscopy and X-ray diffraction. The detailed experimental parameters are described elsewhere [GAL 10d].

3.3.2. *Tribologically-induced molecular changes*

3.3.2.1. *Tribologically-induced molecular orientation right at the surface*

The wear scar of the samples was initially investigated using scanning electron microscopy. The samples were prepared for SEM using gold sputtering. In this method, a blurred appearance of the surface was observed, but not much of the molecular orientation could be revealed. Nevertheless one detail is worth to mentioning. Even with the naked eye, on all samples a prominent blister was observed (Figure 3.26), that can also be seen in the profiles of the wear scars (Figures 3.19-3.21). This prominent blister was still connected with the underlying material on one side, while it was loose on the other.

A comparison with the loading parameters revealed that this blister always occurred in the area where the highest loading, equal to the combination of load and relative velocity, was found. In accordance with the direction of motion of the translating wheel at the part of the cycle, when the highest loads occur, a deformation mechanism shown in Figure 3.27 is proposed. A similar wear behavior is often observed in the wear of elastomers. For rubbers, it is known as “roll formation” [STA 01].

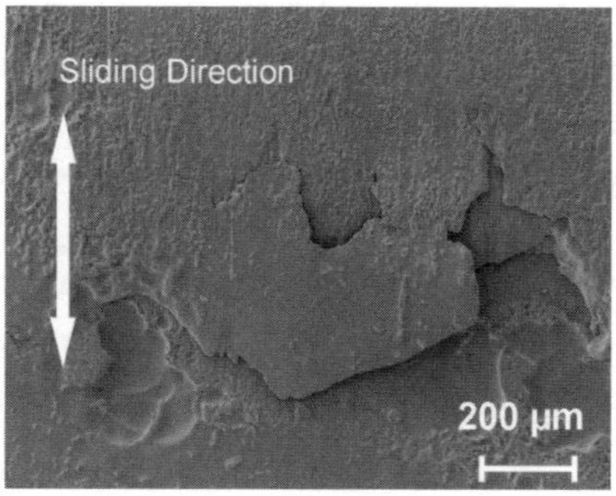

Figure 3.26. *Typical appearance of the blister (top view) (reprinted from [GAL 10d] © 2010, with permission from Springer)*

To investigate the degree of crystallinity right at the surface, Raman spectroscopy was used. The penetration depth of the Raman signal is only 3 µm, meaning the signal derives from that depth or less [ZER 89]. Using this technique, reduced cystallinity at the surface was observed after tribological loading, as can be qualitatively seen for the increase of all amorphous peeks in a typical spectrum of the wear scar, compared to pristine, unloaded material, reported into detail elsewhere [GAL 10d].

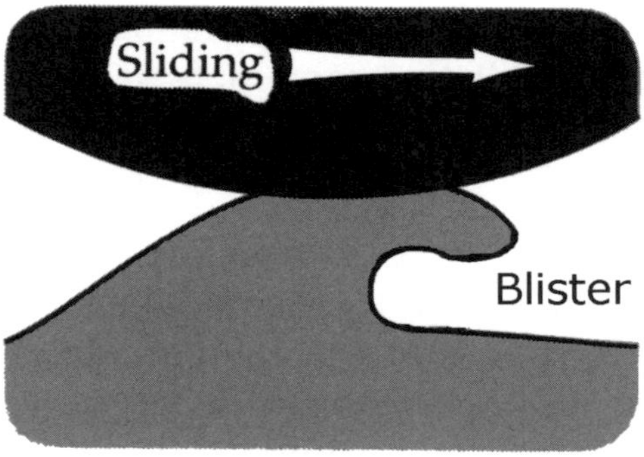

Figure 3.27. *Schematic appearance of the blister (side view)*

The reduction in crystallinity changes along the wear scar, which can easily be explained, as the loading does as well. Generally, the ordered crystalline structure within the polymer is destroyed and transferred material to the CoCrMo wheels was observed [GAL 10c], meaning the adhesive interaction between the bearing-materials must be quite high during such tribological loading. The observation that this layer was totally removed during the etching process for the SEM supports the observation from the Raman spectroscopy of a rather amorphous layer at the surface.

3.3.2.2. *Tribologically-induced molecular orientation in the wear scar under the surface at a depth of about 5-20 μm*

To investigate the zone at a certain depth, the samples were etched using a procedure described and investigated in detail in [NAY 83]. The amorphous part of the UHMWPE is selectively taken away by this etching process. Afterwards, a strong micro-fibrillar orientation was observed using SEM microscopy, as shown in Figure 3.29.

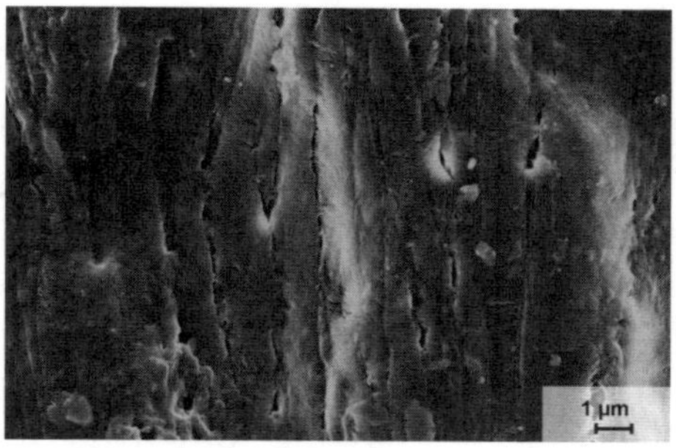

Figure 3.28. *Microfibrils (reprinted from [GAL 10d] © 2010, with permission from Springer)*

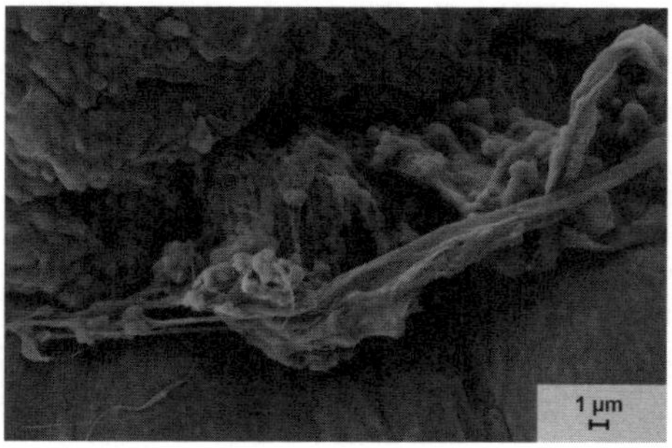

Figure 3.29. *Typical Debris (reprinted from [GAL 10d] © 2010, with permission from Springer)*

A similar, friction-induced orientation of the molecules in the zone close to, but not right at, the surface was previously observed in [DHA 01, WAN 98]. To generate such a microfibrillar structure, high stresses have to occur in this depth of the sample. Also in the area of the blister, the orientation of these microfibrils is most pronounced. It was previously shown [GAL 10c], that the blister is generated in an area that is loaded twice during one cycle with high loads, in combination with high relative velocities (high wheel slip).

The high friction energy can increase the temperature locally, see section 3.1, which facilitates the orientation of the polymer. During this loading, the stress is not equally applied between the molecules due to their different chain length or degree of stretching. Chain scission occurs. This leads to further stress concentrations on the remaining molecules and leaves highly reactive free radicals [PET 71]. This finding is confirmed by a DSC measurement of some debris which shows higher crystallinity. From this measurement it could be seen that the higher crystallinity is not only an effect due to drawing, but a higher crystallinity is due to a decrease in molecular weight. Additionally, the melting point of the debris was lower compared to the pristine material. This shows a degradation of the molecular weight and can explain why the highest degree of oxidation is always found underneath the surface by other authors [PLU 00].

The crystallinity right at the very surface (1-3 µm), which was determined using Raman spectroscopy, decreases in contrast to that of the debris that usually was 10 or more µm thick lamellae. This shows that the large majority of wear particles do not derive from the very top of the bearing surface, but are generated when micro-fibrils become loose. This is affirmed by the appearance of a typical wear particle in Figure 3.28. The high deformation and plastic flow necessary to induce the formation of microfibrils and the blister is only possible because of the tie-molecules that connect the different crystallites. The greater ultimate strain of UHMWPE, compared to lower molecular weight polyethylene, is one explanation for the success of this version of polyethylene in artificial joints. The high ultimate strain of UHMWPE is even in the range of some rubbers and explains why the "blister" resembles the wear manifestation "roll formation".

Nevertheless, high stretching involves a degradation of the bonds between adjacent microfibrils and the material below. Therefore, microfibrils are only sparsely connected to the underlying material and are a precursor to wear. These observations also explain why wear is much higher when multidirectional sliding is applied. Here, shear stresses additionally occur in between adjacent lamellae [BRA 96]. In the described unidirectional wheel-on-flat tests, gravimetric measurements were unable to reveal a reliable weight loss in comparison to unloaded samples put in water at the corresponding temperature. With the wear rate that low, most of the contact deformation reported in Figures 3.19-3.21 is due to creep.

3.3.2.3. *Tribologically induced molecular orientation in the wear scar under the surface at a depth of more than 20 μm*

In an orientation measurement using X-ray diffraction in Bragg-Bretano geometry, neither the amorphous film in the first 1-3 μm nor the orientation of the microfibrils that extend to a depth of 10-20 μm under the surface could be determined. Instead, the orientation of the a- and b-axis in the depth of the samples of the orthorhombic crystal was obtained.

The orientation of the c-axis is perpendicular to this and corresponds directly to the direction of lamellar alignment. The prevailing crystal orientation is shown in Figure 3.30. The deeper the wear track, the more pronounced the degree of orientation [GAL 10a].

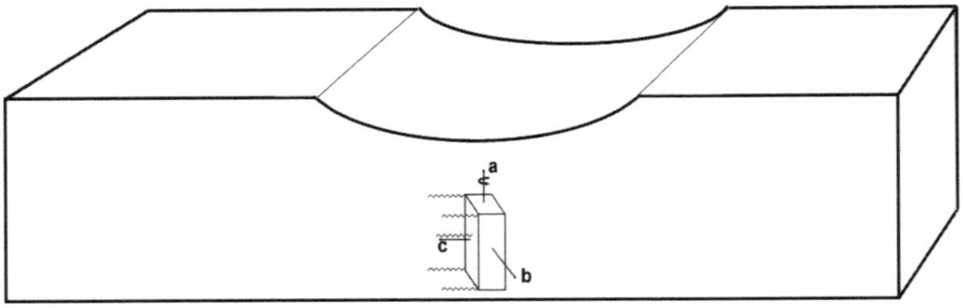

Figure 3.30. *Orientation of the orthorhombic crystal structure under the wear scar after tribological loading*

This orientation in the depth of the sample was found to be the same as after reciprocal unidirectional compression fatigue, tested as described elsewhere [GAL 07]. Therefore the same mechanisms are believed to occur in that zone under frictional as well as under reciprocal compression loading. Thus the model proposed by [GAL 92] is favored for unidirectional loading.

This model involves a reorientation and rearrangement of crystals under compression and results in a new long-range order. In this rearrangement, the chain-slip processes in the crystals act in combination with the shear deformation of the amorphous phase. The observed X-ray pattern of the tribologically tested samples proves that this deformation occurs under physiologically relevant loads.

3.3.2.4. *Overview of the different molecular deformation mechanisms and zones in the UHMWPE under tribological rolling-sliding motion*

In Figure 3.31, an overview of the different crystal orientation in the wear scar and the corresponding estimated depths scale is given.

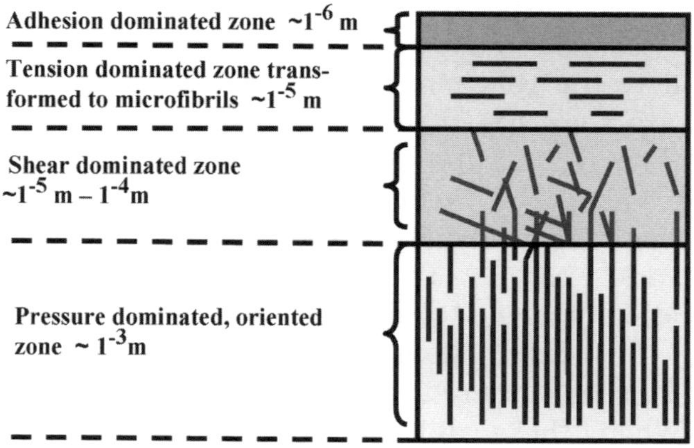

Figure 3.31. *Model of the deformation in the UHMWPE under friction (reprinted from [GAL 10d] © 2010, with permission from Springer)*

Surprisingly, when the orientation of the crystals in the different zones of UHMWPE after tribological loading is examined, it very much resembles the orientation of collagen fibrils in human articular cartilage, observed by Benninghoff [BEN 25] and later confirmed by many others. This is shown in Figure 3.32.

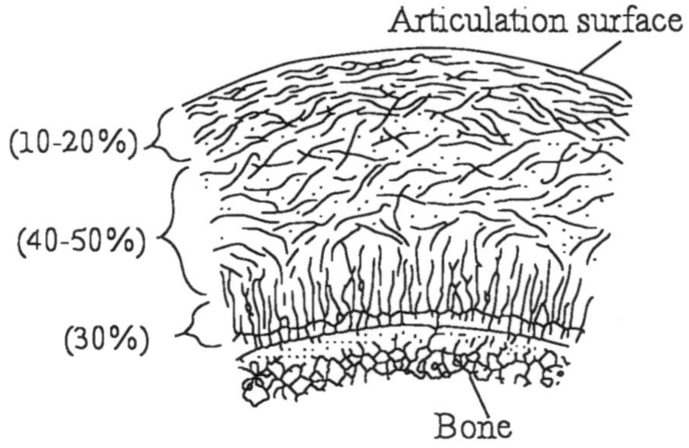

Figure 3.32. *Orientation of the collagen fibrils in human cartilage according to Benninghoff's cartilage model [ROS 03]*

The amorphous layer at the very top of the UHMWPE sample has no similarity to natural cartilage, but surprisingly the collagen fibrils and the polymer lamellae underneath show the same orientation. The problem in the present test is that the normal load and relative velocity changed along the wear scar. The degree and depth distribution of microstructural deformation change along the wear track, as do the loading parameters; further investigation is needed under constant loading conditions to clarify the interactions. Nevertheless the general mechanisms and orientation are the same along the wear track. When these different zones of deformation are compared with the general stresses under a slider (Figure 3.33), it is obvious how the microstructural changes were generated.

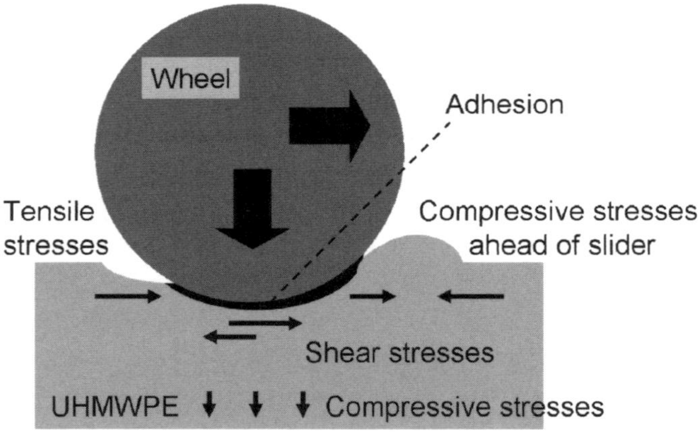

Figure 3.33. *Distribution of stresses under a slider (reprinted from [GAL 10d] © 2010, with permission from Springer)*

At the surface, adhesion and therefore high shear stresses destroy the crystalline structure of the polymer. A slider and especially a slipping wheel also evoke tangential forces. These stresses lead to highly oriented microfibrils. These tensile stresses appear to be more effective in changing the microstructure and are therefore more detrimental than the compression stress ahead of the slider and the shear stresses underneath the slider. The tensile stresses may also induce crazes under non-unidirectional loading and therefore wear particles. Under the orientated layer and a barely oriented transition zone reciprocal compression stresses mainly occur, that lead to an orientation perpendicular to the one close to the surface. In between is a zone where high shear is applied according to the Hertzian model.

Due to the friction force at the surface the maximum shear occurrence is not as deep as was proposed by the Herzian equations. Nevertheless it occurs at a certain depth were the polymer can hardly adapt and is even weakened due to the transition between the microfibrillar orientation and the compression induced orientation. The good agreement with results, reported for other semi-crystalline polymers such as high density polyethylene [BAR 02], polytetraflourpolyethylene [STA 01] or polyamide [WEI 94 MAR 09] suggests that the observed mechanisms are in fact generic for many linear semi-crystalline polymers. The prevalence of the different particular mechanisms instead depends on micro structural properties such as molecular weight or chemical composition and the loading conditions.

3.4. Importance for artificial knee joints

UHMWPE has been extensively tested against CoCrMo at parameters in the physiological range of knee joints. During friction a significant temperature rise was observed under the surface of the polymer. The temperature rise is linearly related to the product of normal load, relative velocity and the coefficient of friction. The measured fraction of heat dissipated within the polyethylene was 0.07. This is twice the amount that was previously used to calculate the temperature rise in artificial knee joints *in vivo* [SAW 03].

Therefore, the theoretical approach in literature by which a temperature rise of 9 K was determined is an underestimation.

It should also be noted, that the influence of the thermal diffusivity of the counterpart materials is high [IMA 04, LIA 99, LU 97]. Therefore materials with a higher thermal diffusivity might show a better performance in artificial joints.

Nevertheless, a temperature rise of some degrees was shown to affect mechanical properties and creep behavior. For example, in a tribological test the same deformation is reached after 3 million cycles at 23°C and 1 million cycles at 45°C (Figures 3.19 and 3.21). Wear is also linked to the applied energy. In tests with the self-made tribological device the wear manifestation "blister" could to be related to the product of normal load and relative velocity. Miura *et al.* [MIU 02] correlated the applied energy to wear, as does Lancaster's general PV-approach "pressure multiplied by velocity" for polymers [LEW 64].

In knee implants, two general concepts compete with each other: a design with a high conformity and a less congruent, more natural round-on-flat design [KOH 99]. It is often argued that a highly congruent design is favorable because it leads to lower contact pressure [PET 99], surprisingly, higher wear was reported with higher congruency [ESS 03]. One reason might be that high congruency in the sagittal plane hinders the physiological roll-back of the femur on the tibia and the tibia component is exposed to higher shear forces that are even transmitted to the fixation [SAT 99, ROS 03]. Another reason follows from the present results, because with such a design the relative velocity between both bodies is higher.

It has been shown here, that not only the load, but also the relative velocity and the friction coefficient must be taken into account. The friction energy is crucial. This brings the advantages of low contact pressure into question. In addition, it is shown that with higher relative velocity the coefficient of friction increases under dry friction giving an even higher product of the three influencing parameters. Therefore, in the best case rolling between the UHMWPE inlay and its counterpart is realized at the points of the walking cycle where high loads apply on the knee joint. The energy input is low and distributed on different contact points. The present findings support the concerns against more-constrained knee prostheses that were shown to generate more heat than less-constrained prostheses [PRI 05].

A new approach is the full imitation of the natural knee that allows physiological motion, especially if the ligaments are intact. In the natural knee the sagittal tibial contours are convexly curved in the lateral compartment, but concavely curved in the medial one. The imitation of this shape allows a high degree of rolling in the joint, when the highest loads apply on the joint during walking [NAE 08]. If the friction energy is taken as an indication of wear, different designs of knee joints could be compared by applying the finite element method.

It can be argued that boundary lubrication occurs *in vivo* [JAL 01, WAL 96], but dry conditions should be taken into account to be on the safe side. Generally, it is hard to tell which further implications might arise from a temperature rise *in vivo*. For example, at elevated temperatures the lubrication *in vivo* can also be affected. The viscosity changes at higher temperatures due to the non-Newtonian properties of the synovial fluid [LIA 99]. At a certain temperature, proteins can even decompose [LIA 03] and the modified proteins become sufficiently chemically active to degrade the UHMWPE and accelerate wear [CHA 98]. An indication that elevated temperatures contribute to wear *in vivo* is the yellowing effect, sometimes found on certain spots of the inlay [YOU 98]. This oxidation related damage is one of the major problems in long-term use and accelerates at higher temperatures [DAV 87].

Additionally, some conclusions for the testing of artificial knee joints can be derived from the findings. It was shown that higher temperatures accelerate the deformation. Section 3.2 showed that the temperature rise is suppressed by cooling with the help of a large amount of liquid. Usually knee simulators are run within several hundred milliliters of liquid. In the natural knee only 2-4 ml of synovia is found [THU 01] and the temperature rises even in natural joints during walking [TEP 85].

Usually simulated knee joints are tested in several hundred ml of liquid. In the natural knee only 2-4 ml synovia are found [THU 01] and this volume may be even less in case of artificial knees when the joint capsule is damaged. Although a value of the heat capacity of synovia could not be found, the comparison with blood (3.59 J/gK [BLA 00] proves that 2-4 ml cannot dissipate all friction energy. That means, if the results from knee simulator tests at 37°C are transferred to artificial joints *in vivo*, we have to take care not to overestimate the performance. On the other hand it might be possible to accelerate *in vitro* tests by increasing the testing temperature. It has been shown that about the same amount of deformation occurs at 45°C after 1,500,000 cycles, as is found after 3,000,000 cycles at 37°C.

The importance of the applied energy and its relation to the temperature rise also offers a simple and direct possibility of identifying highly stressed spots on an inlay. The temperature increase of an artificial knee joint tested *in vitro* can be measured with a thermal camera. This would be a much easier and faster method to compare different designs than conventional wear testing.

Finally, a model was proposed which reflects the deformation mechanisms during friction, see section 3.3.2.4. From this model two improvements can be proposed for the processing of UHMWPE. Better wear behavior might be achieved by pre-conditioning the samples with exposure to compression loads at a temperature close to the melting point. This is meant to create the required orientation to reduce creep and shear within the material during tribological loading. Ohta *et al.* already tested this approach with promising results [OHT 99]. It might even be useful to pay attention to the inherent orientation from the sintering process under pressure, when the implants are machined. At this point further research is necessary in the future. Additionally, a certain degree of cross-linking is beneficial to restrict the huge plastic deformation close to the surface and reduce the molecular orientation. A high molecular mismatch between certain zones is detrimental as it weakens the strength of the polymer. This approach was proven to be beneficial due to the decreased wear behavior of cross-linked material [MUR 99].

As a conclusion, the findings show the need to further identify and validate the temperature in artificial knee joints. This will enable the prediction of long-term performance of implants more precisely.

3.5. Acknowledgments

The authors would like to thank Elsevier, Springer and Shaker for their kind permission to reuse previously published material. The authors also wish to express their sincere gratitude to Professor Peter Adam for all the interesting discussions.

3.6. Bibliography

[ACH 01] ACHENBACH M., FRANK E., "Reibung von Elastomeren", *Tribologie + Schmierungstechnik*, vol. 48, pp. 43-47, 2001.

[BAR 92] BARTCZAK Z., "Macromolecules Evolution of the crystalline texture of high-density polyethylene during uniaxial compression", *Macromolecules*, vol. 25, pp. 4692-4704, 1992.

[BAR 95] BARTCZAK Z., ARGON A.S., COHEN R.E., "Deformation mechanisms and plastic resistance in single-crystal-textured high-density polyethylene", *Macromolecules*, vol. 25, pp. 5036-5053, 1995.

[BAR 02] BARTCZAK Z., "Deformation of high-density polyethylene produced by rolling with side constraints. I. Orientation behavior", *Journal of Applied Polymer Science*, vol. 86, pp. 1396–1404, 2002.

[BAR 79] BARTON, S.J., CHERRY, B.W., "Predicting the creep rupture life on polyethylene pipe", *Polymer Eng. Sci.*, vol. 19 no. 8, pp. 590-595, 1979.

[BEN 25] BENNINGHOFF A., "Form und Bau der Gelenkknorpel in ihren Beziehungen zur Funktion", *Z. Zellforsch Mikrosk Anat*, vol. 2, p. 814, 1925.

[BER 07] BERGER C., BOCKENHEIMER A., "Lebensdauerabschätzung an Kunststoffrohren mittels Zeitstand-Innendruckversuch", *Materialwissen und Werkstofftech*, vol. 38, pp. 418-422, 2007.

[BER 01] BERGMANN G., GRAICHEN F., ROHLMANN A., VERDONSCHOT N, VAN LENTHE GH., "Frictional heating of total hip implants. Part 1: measurements in patients", *J Biomech*, vol. 34, pp. 421-428, 2001.

[BHU 01] BHUSHAN B., *Modern Tribology Handbook: Vol. 1-Principles of Tribology*, CRC Press, 2001.

[BLA 00] BLAKE A.S.T., PETLEY G.W., DEAKIN C.D., "Effects of changes in packed cell volume on the specific heat capacity of blood: implications for studies measuring heat exchange in extracorporeal circuits", *BJA Brit. J. Anaesthesia*, vol. 84, pp. 28-32, 2000.

[BLO 37] BLOK H., "Theoretical study of temperature rise at surfaces of actual contact under oiliness lubricating conditions", *Proc. General Discussion on Lubrication Inst. Mech. Eng*, vol. 2, pp. 222-235, 1937.

[BOW 94] BOWDEN P.B., YOUNG R.J., "Deformation mechanisms in crystalline polymers", *J. Mat. Sci.*, vol. 29, pp. 294-323, 1994.

[BRA 96] BRAGDON C.R., O'CONNOR O.D., LOWENSTEIN J.D., JASTY M., SYNIUTA W.D., "The importance of multidirectional motion on the wear of polyethylene", *Proc. Instn. Mech. Engrs. Part H*, vol. 210, pp. 157-166, 1996.

[BRI 81] BRISCOE B.J., "Friction and wear of organic solids and the adhesion model of friction", *Phil. Magazine*, vol. A43, pp. 511-527, 1981.

[BQS 07] BQS Bundesgeschäftsstelle Qualitätssicherung GmbH. BQS-Qualitätsreport, 2007.

[CAR 59] CARSLAW H.S., JAEGER J.C., *Conduction of Heat in Solids*, Clarendon, Oxford, 1959.

[CHA 98] CHANDRASEKARAN M., WEI L.Y., VENKATESHWARAN K.K., BATCHELOR A.W., LOH N.L., "Tribology of UHMWPE tested against a stainless steel counterface in unidirectional sliding in presence of model synovial fluids: Part 1", *Wear*, vol. 223, pp. 13-21, 1998.

[CHO 04] CHO H.J., WEI W.J., KAO H.C., CHENG C.K., "Wear behavior of UHMWPE sliding on artificial hip arthroplasty materials", *Materials Chem. Phys.*, vol. 88, pp. 9-16, 2004.

[CRI 89] CRIST, B., FISHER, C.J., HOWARD, P.R., "Mechanical properties of model polyethylenes: tensile elastic modulus and yield stress", *Macromolecules*, vol. 22, pp. 1709-1718, 1989.

[DAS 03] DASARI A., MISRA RDK., "On the strain rate sensitivity of high density polyethylene and polypropylenes", *J. Mater. Sci. Eng.*, vol. A1, pp. 356-371, 2003.

[DAV 87] DAVIDSON J.A., SCHWARTZ G., "Wear, Creep, and Frictional Heat of Femoral Implant Articulating Surfaces and the Effect on Long-Term Performance-Part 1, Review", *J. Biomed. Mat. Res.*, vol. 21, pp. 261-285, 1987.

[DHA 01] DHARMASTITI R., BARTON D.C., FISHER J., EDIDIN A., KURTZ S.M., "The wear of oriented UHMWPE under isotropically rough and scratched counterface test conditions", *Bio-Medical Mat. and Eng.*, vol. 11, pp. 241-256, 2001.

[DON 07] DONG G.N., HUA M., LI J., CHUAH K.B., "Temperature field and wear prediction for UHMWPE acetabular cup with assumed rectangular surface texture", *Materials and Design*, vol. 28, pp. 2402-2416, 2007.

[DOW 87] DOWSON D., TAHERI S., WALLBRIDGE N.C., "The role of counterface imperfections in the wear of polyethylene", *Wear*, vol. 119, pp. 277-293, 1987.

[DEN 97] DENG M., LATOUR R.A., OGALE A.A., SHALABY S.W., "Study of creep behavior of ultra-high-molecular-weight polyethylene systems", *J. Biomed. Mat. Res., Part A*, vol. 40, pp. 214-223, 1997.

[EDI 99] EDIDIN A.A., PRUITT L., JEWETT C.W., CRANE D.J., ROBERTS D., KURTZ, S.M., "Plasticity induced damage layer is a precursor to wear in radiation-cross-linked UHMWPE acetabular components for total hip replacement", *J. Arthroplasty*, vol. 14, pp. 616-27, 1999.

[ESS 03] ESSNER A., KLEIN R., BUSHELOW M., WANG A., KVITNITSKY M., MAHONEY O., "The effect of sagittal conformity on the knee", *Wear*, vol. 255, pp. 1085-1092, 2003.

[EYR 36] EYRING H., "Viscosity, plasticity, and diffusion as examples of absolute reaction rates", *J. Chem. Phy.*, vol. 4, pp. 283–291, 1936.

[FOT 78] FOTHERINGHAM D.G., CHERRY B.W., "The role of recovery forces in the deformation of linear polyethylene", *J. Mater. Sci.*, vol. 13, p. 951–964, 1978.

[FRI 83] FRIEDRICH K., "Crazes and shear bands in semi-crystalline thermoplastics", *Adv. Polym. Sci.*, vol. 52, pp. 225-274, 1983.

[FUR 03] FURUKAMA J., *Physical Chemistry of Polymer Rheology*, Heidelberg - Springer, 2003.

[GAL 92] GALESKI A., BARTCZAK Z., ARGON A.S., COHEN R.E., "Morphological alterations during texture-producing plastic plane strain compression of high-density polyethylene", *Macromolecules*, vol. 25, pp. 5705-5718, 1992.

[GAL 07] GALETZ M.C., GOETZ C., ADAM P., GLATZEL U., "Hysteretic heating during cyclic loading of medical grade ultra high molecular weight polyethylene", *Advanced Engineering Materials*, vol. 9, pp. 1089-1096, 2007.

[GAL 09a] GALETZ M.C., UTH T., WIMMER M.A., ADAM P., GLATZEL U., "Determination of the temperature rise within UHMWPE tibial components during tribological loading", *Acta Biomaterialia*, vol. 6, pp. 552-562, 2009.

[GAL 09b] GALETZ M.C., *Influence of Temperature on Friction and Deformation Mechanisms in Artificial Knee Joints*, Shaker, Aachen, 2009.

[GAL 10a] GALETZ M.C., GLATZEL U., "An activated energy approach for accelerated testing of the deformation of UHMWPE in artificial joints", *Journal of the Mechanical Behavior of Biomedical Materials*, vol. 3, pp. 331-338, 2010.

[GAL 10b] GALETZ M.C., FLEISCHMANN E., KONRAD C., SCHÜTZ A., GLATZEL U., "Abrasion resistance of oxidized zirconium in comparison with CoCrMo and titanium nitride coatings for artificial knee joints", *Journal of Biomedical Materials Research, Part B - Applied Biomaterials*, vol. 93-1, pp. 244-251, 2010.

[GAL 10c] GALETZ M., DIETEL S., THEILE B., GLATZEL U., "Potential for adhesive wear in friction couples of UHMWPE running against oxidized zirconium, titanium nitride coatings and cobalt-chromium alloys", *Journal of Biomedical Materials Research, Part B - Applied Biomaterials*, vol. 93-2, pp. 468-475, 2010.

[GAL 10d] GALETZ M.C., GLATZEL U., "Molecular deformation mechanisms in UHMWPE during tribological loading in artificial joints", *Tribology Letters*, vol. 38, pp. 1-13, 2010.

[GRE 60] GREENWOOD J.A., MINCHALL H., TABOR D., "Hysteresis losses in rolling and sliding friction", *Proc. Royal. Soc.*, vol. A259, pp. 480-507, 1960.

[HOO 83] HOOD R.W., WRIGHT T.M., BURSTEIN A.H., "Retrieval analysis of total knee prostheses: A method and its application to 48 total condylar prostheses", *J. Biomed. Mat. Res.*, vol. 17, pp. 829-842, 1983.

[IMA 04] IMADO K., MIURA A., NAGATOSHI M., KIDO Y., MIYAGAWA H., HIGAKI H., "A study of contact temperature due to frictional heating of UHMWPE", *Tribol. Lett.*, vol. 16, pp. 265-273, 2004.

[JAL 01] JALALI-VAHID D., JAGATIA M., JIN Z.M., DOWSON D., "Prediction of lubricating film thickness in UHMWPE hip joint replacements", *J. Biomech.*, 34, pp. 261–266, 2001.

[KAU 72] KAUSCH H.H., in "Advances in Polymer Science and Engineering", edited by PAE P.A, New York, pp. 207, 1972.

[KEN 00] KENNEDY F.E,. CURRIER J.H., PLUMET S., DUDA J.L., GESTWICK D.P., COLLIER JP, CURRIER B.H., "Contact fatigue failure of ultra-high molecular weight polyethylene bearing components of knee prostheses", *J. Tribol.*, vol. 122, pp. 332-340, 2000.

[KOH 99] KOHN D., RUPP S., "Alloarthroplastik des Kniegelenkes", *Der Orthopäde*, vol. 28, pp. 975-995, 1999.

[KRA83] KRAGELSKI I., DOBYCIN N., KOMBALOV V, *Grundlagen der Berechnung von Reibung und Verschleiß*, Hanser, 1983.

[KUR 98] KURTZ S.M., PRUITT L., JEWETT C.W., CRAWFORD R.P., CRANE D.J., EDIDIN A.A., "The yielding, plastic flow, and fracture behavior of ultra-high molecular weight polyethylene used in total joint replacements", *Biomaterials*, vol. 19, pp. 1989-2003, 1998.

[KUR 99] KURTZ S.M., MURATOLGU O.K., EVANS M., EDIDIN A.A.. "Advances in the processing, sterilization, and crosslinking of ultra-high molecular weight polyethylene for total joint arthroplasty", *Biomaterials*, vol. 20, pp. 1659-1688, 1999.

[KUR 02] KURTZ S.M., VILLARAGA M.L., HERR M.P., BERGSTRÖM J.S., RIMNAC C.M., EDIDIN A.A., "Thermomechanical behaviour of virgin and highly crosslinked ultra-high molecular weight polyethylene used in total joint replacements", *Biomat.*, vol. 23, pp. 3681-3697, 2002.

[KUR 04] KURTZ S.M., *The UHMWPE Handbook*, Elsevier Academic Press, SanDiego, 2004.

[LAN 79] LANCASTER, J.K., EVANS, D.C., "The wear of polymers", in *Treatise on Materials Science and Technology*, SCOTT D (Ed.), Academic Press, 1979

[LEW 64] LEWIS R.B., "Predicting the wear of sliding plastic surfaces", *Mech. Eng.*, vol.86, pp. 32-40, 1964.

[LIA 99] LIAO Y.S., BENYA P.D., MCKELLOP H.A., "Effect of protein lubrication on the wear properties of materials for prosthetic joints", *Appl. Biomat.*, vol. 48, p. 465-473, 1999.

[LIA 03] LIAO Y.S., MCKELLOP H., LU Z., CAMPBELL P., BENYA P., "The effect of frictional heating and forced cooling on the serum lubricant and wear of UHMW polyethylene cups against cobalt–chromium and zirconia balls", *Biomaterials*, vol. 24, p. 3047-59, 2003.

[LIN 94] LIN L., ARGON A.S,. "Structure and plastic deformation of polyethylene", *Journal of Materials Science*, vol. 29, pp. 294-223, 1994.

[LU 97] LU Z., MCKELLOP H., "Frictional heating of bearing materials tested in a hip joint wear simulator", *Proc. Inst. Mech. Eng. Part H: Eng. in Med.*, vol. 211, pp. 101-108, 1997.

[MAR 09] MARCELLAN A., BONDIL O., BOUÉ C., CHATEAUMINOIS A., "Third body effects in the wear of polyamide: micro-mechanisms and wear particles analysis", *Wear*, vol. 266, pp. 1013-1020, 2009.

[MAR 91] MARCUS K., BALL A., ALLEN C., "The effect of grinding direction on the nature of the transfer layer formed during the sliding wear of ultrahigh molecular weight polyethylene against stainless steel", *Wear*, vol. 151, pp. 2323-336, 1991.

[MCK 81] MCKELLOP H., CLARKE I., MARKOLF K, AMSTUTZ H., "Friction and wear properties of polymer, metal and ceramic prosthetic joint materials evaluated on a multi-channel screening device", *J. Biomed. Mater Res.*, vol. 15, pp. 619-653, 1981.

[MCK 07] MCKELLOP H.A., "The lexicon of polyethylene wear in artificial joints", *Biomaterials*, vol. 28, pp. 5049-505, 2007.

[MIU 02] MIURA H., "Prediction of total knee arthroplasty polyethylene wear using the wear index", *J. Arthroplasty*, vol. 17, pp. 760-766, 2002.

[MUR 99] MURATOGLU O.K., BRAGDON C.R., O'CONNOR D.O., JASTY M., HARRIS H.H., GUL R., MCGARRY F., "Unified wear model for highly crosslinked ultra-high molecular weight polyethylenes (UHMWPE)", *Biomat.,* vol. 20, pp. 1463-1470, 1999.

[NAE 08] NÄGERL H., FROSCH K.H., WACHOWSKI M.M., DUMONT C., ABICHT C., ADAM P., KUBEIN-MEESENBURG D, "A novel total knee replacement by rolling articulating surfaces. In vivo functional measurements and tests", *Acta. of Bioengi. and Biomech.*, vol. 10, pp. 55-60, 2008

[NAY 83] NAYLOR K.L., PHILLIPS P.J., "Optimization of permanganic etching of polyethylenes for scanning electron microscopy", *J. Polym. Sci.*, vol. 21, pp. 2011-2026, 1983.

[OHT 99] OHTA M., HYON S.H., OKA M., TSUTSUMI S., "Wear resistance of lightly cross-linked ultrahigh-molecular-weight polyethylene crystallized from the melt under uniaxial compression", *Wear*, vol. 225, pp. 312-318, 1999.

[OLE 03] OLEINIK E.F., "Plasticity of semicrystalline flexible-chain polymers at the microscopic and mesoscopic levels", *Polymer science Series C*, vol. 45, pp.17-117, 2003.

[PAT 78] PATERMANN J., SCHULTZ J.M., "Lamellar separation during the deformation of high-density polyethylene", *J. Mat. Sci.*, vol. 13, p. 50, 1978.

[PAU 76] PAUL J.P., "Loading on normal hip and knee joints and on joint replacements" in *Advances in Artificial Hip and Knee Joint Technology*, ed. SCHALDACH M, HOHMANN D, Springer, Berlin, 1976

[PET 71] PETERLIN A., "Molecular model of drawing polyethylene and polypropylene", *J. Mat. Sci.*, vol. 6-6, pp. 1573-4803, 1971.

[PET 66] PETERSON J.M., LINDENMEYER PH, "Screw dislocations in anisotropic media", *Journal of Applied Physics*, vol. 37, pp. 4051, 1966.

[PET 99] PETTY W., MILLER G.J., BARTEL D.L., WRIGHT T.M., BURSTEIN A.H., "Total knee contact pressures: the effect of congruity and alignment", *Medscape General Med*, vol. 1, pp. 1-5, 1999.

[PLU 00] PLUMET S., DUDA J.L., GESTWICK D.P., COLLIER J.P., CURRIER B.H., "Contact fatigue failure of ultra-high molecular weight polyethylene bearing components of knee prostheses", *J. Tribology*, 122, pp. 332-340, 2000.

[POO 72] POOLEY C.M, TABOR D., "Friction and molecular structure. the behaviour of some thermoplastics", *Proc. the Royal Society of London. Series A, Mathematical and Physical Sciences*, vol. 329, pp. 251-274, 1972.

[POP 75] POPE D.P., KELLER A., "Deformation of oriented polyethylene", *J. Polymer Sci.: Polymer Physics Edition*, vol. 13, 533-566, 1975.

[PRE 66] PREDECKI P., STATTON W.O., "Dislocations caused by chain ends in crystalline polymers", *J. Appl. Phys.*, vol. 37, pp. 4053-4059, 1966.

[PRI 05] PRITCHETT J.W., "Heat generated by knee protheses", *Clin. Orthop. Relat. Res.*, vol. 442, pp. 195-198, 2005.

[ROS 03] ROSENBERG C.H.T. "Verschleißverhalten von Polyäthylen beim künstlichen Kniegelenkersatz", Online-Dissertation, Aachen, 2003.

[SAI 01] SAIKKO V., AHLROOS T., CALONIUS O., "A three-axis knee wear simulator with ball-on-flat contact", *Wear*, vol. 249, pp. 310-315, 2001.

[SAM 04] SAMYN P., SCHOUKENS G., "Calculation and significance of the maximum polymer surface temperature T* in reciprocating cylinder-on-plate sliding", *Polym. Eng. Sci*,, vol. 48, p. 774-785, 2004.

[SAT 99] SATHASIVAM S., WALKER PS., "The conflicting requirements of laxity and conformity in total knee replacement", *J. Biomechanics*, vol. 32, 1999, p. 239-247.

[SAW 03] SAWYER W.G., HAMILTON M.A., FREGLY B.J., BANKS S.A., "Temperature modeling in a total knee joint replacement using patient-specific kinematics", *Tribol. Lett.*, 15, pp. 343–351, 2003.

[SCH 99] SCHMALZRIED T., "Current concepts review - wear in total hip and knee replacements", *J. Bone and Joint Surgery*, vol. 81, pp. 115-136, 1999.

[SCH 05] SCHWENKE T., BORGSTEDE L.L., SCHNEIDER E., ANDRIACCHI T.P., WIMMER M.A., "The influence of slip velocity on wear of total knee arthroplasty", *Wear*, vol. 259, p. 926-932, 2005.

[SEQ 94] SÉGUÉLA R., DARRAS O., "Phenomenological aspects of the double yield of polyethylene and related copolymers under tensile loading", *J. Mater. Sci.*, vol. 29, pp. 5342-5352, 1994.

[SEQ 02] SÉGUÉLA R., "Dislocation approach to the plastic deformation of semicrystalline polymers: kinetic aspects for polyethylene and polypropylene", *J. Polym. Sci. Polym. Phys.* Vol. 40, pp. 593-601, 2002.

[SHA 76] SHADRAKE L.G., GUIU F., "Dislocations in polyethylene crystals: Line energies and deformation modes", *Philos. Mag.,* vol. 34, p. 565-581, 1976.

[STA 01] STACHOWIAK G.W., BATCHELOR A.W., *Engineering Tribology*, Butterworth Heinemann,Woburn, 2001

[SWE 08] Swedish Knee Arthroplasty Register: Annual Report, Lund, Sweden, 2008.

[TAN 77] TANAKA K., MIYATA T., "Study on the friction and transfer of semicrystalline polymers", *Wear*, vol. 41, p. 383, 1977.

[TEP 85] TEPIC S., MACIROWSKI T., MANN R.W., "Experiamental temperature rise in human hip joint in vitro in simulated walking", *J. Orthop. Res.*, vol. 3, pp. 516-520, 1985.

[THU 01] THUMB N., BRÖLL H., CZURDA R., *Praktische Rheumatologie*, Springer 2001.

[TRU 84] TRUSS R.W., CLARKE P.L., DUCKETT R.A., WARD I.M, "The dependence of yield behavior on temperature, pressure, and strain rate for linear polyethylenes of different molecular weight and morphology", *J. Polym. Sci, Polym. Phys.*, vol. 22, pp. 191-209, 1984.

[TUR 05] TURELL M.E., FRIEDLAENDER G.E., WANG A., THORNHILL T.S., BELLARE A., "The effect of counterface roughness on the wear of UHMWPE for rectangular wear paths", *Wear*, vol. 259, pp. 984-991, 2005.

[UET 85] UETZ H., WIEDEMEYER J., *Tribologie der Polymere*, Carl Hanser, München, Wien, 1985.

[URB 01] URBAN J.A., GARVIN K.L., BOESE C.K., BRYSON L., PEDERSEN D.R., CALLAGHAN J.J., MILLER R.K., "Ceramic-on-polyethylene bearing surfaces in total hip arthroplasty. Seventeen to twenty-one-year results", *J. Bone Joint Surg. Am.*, vol. 82, pp. 1688-1694, 2001.

[WAL 96] WALKER P.S., BLUNN G.W., LILLEY P.A., "Wear testing of materials and surfaces for total knee replacement", *J. Biomed. Mat. Res.*, vol. 33, pp. 159-175, 1996.

[WAN 95] WANG A., STARK C., DUMBLETON J.H., "Role of cyclic plastic deformation in the wear of UHMWPE acetabular cups", *Biomed. Mat. Res.*, vol.29, pp. 619-626, 1995.

[WAN 96] WANG A., STARK C., DUMBLETON J.H., "Mechanistic and morphological origins of ultra-high molecular weight polyethylene wear debris in total joint replacement prostheses", *Proc. Instn. Mech. Eng. Part H: J. Eng. Med.*, vol. 210, pp.141-155, 1996.

[WAN 98] WANG A., EDWARDS B., YAU S.S., POLINENI V.K., ESSNER A., KLEIN R., SUN D.C., STARK C., DUMBLETON J.H., "Orientation Softening as a mechanism of Ultra-high molecular weight polyethylene (UHMWPE) Wear in Artificial Hip and Knee joints", in: *Characterization and Properties of Ultra-High Molecular Weight Polyethylene*, (ed.) GSELL R.A., STEIN H.L., PLOSKONDA J.J., *ASTM*, p. 56-79, 1998.

[WAN 01] WANG A., ESSNER A., KLEIN R., "Effect of contact stress on friction and wear of ultra-high molecular weight polyethylene in total hip replacement", *Proc. Inst. Mech. Eng. Part H: J. Eng. in Med.*, 205, pp. 133-139, 2001.

[WEI 94] WEICK B.L., FUREY M.J., NEWMAN H., KAJDAS C., HELLGETH J.W., "Tribochemical change of nylon 6, 6 and nylon 6, 6/glass rubbed against sapphire", *Tribol. Trans.*, vol. 37, pp. 129–137, 1994.

[WIL 78] WILDING MA. WARD IM., "Tensile creep and recovery in ultrahigh modulus linear polyethylenes", *Polymer*, vol. 9, pp. 969-976, 1978.

[WIE 07] WIELEBA A., "The Mechanism of Tribological Wear of Thermoplastic Materials", *Arch. Civil. Mech. Eng.*, vol. 4, pp. 185-191, 2007.

[WIL 55] WILLIAMS M.L., LANDEL R.F., FERRY J.D., "The temperature dependence of relaxation mechanisms in amorphous polymers and other glass-forming liquids", *J. Amer. Chem. Soc.*, vol. 77, pp. 3701-3707, 1955

[YOU 73] YOUNG R.J., BOWDEN P.B., RITCHIE J.M., RIDER J.G., "Deformation mechanism in oriented high density polyethylene", *J. Mater. Sci.*, vol. 8, pp. 23-36, 1973

[YOU 98] YOUNG T.H., CHENG C.K., LEE Y.M., CHEN L.Y., HAUNG C.H., "Analysis of ultrahigh molecular weight polyethylene failure in artifical knee joints: thermal effect on long-term performance", *J. Biomed. Mat. Res.*, vol. 48, pp. 159-164, 1998

[ZER 89] ZERBI G., GALLINO G., DEL FANTI N., BAINI L., "Structural depth profiling in polyethylene films by multiple internal reflection infra-red spectroscopy", *Polymer*, vol. 30, pp. 2324-2327, 1989.

[ZHO 98] ZHOU H., WILKES GL., "Creep behavior of high density polyethylene films having well defined morphologies of stacked lamellae with and without an observable row nucleated fibril structure", *Polymer*, vol. 39, pp. 3597-3609, 1998.

Chapter 4

Large Capacity Wear Testing

This chapter focuses on wear testing of orthopedic biomaterials by means of a large capacity wear simulator consisting of 100 separate test stations. In particular, the typical situation faced by a biotribologist who needs to determine if a certain material has a wear resistance greater to that of another is discussed. The number of specimens in a sample (n), magnitude of the standard deviation (S.D.) of the wear rate (percentage of the mean value) as a measure of dispersion, and comparison of sample mean wear rates are dealt with.

4.1. Introduction

Since the typical wear test for prosthetic joint materials consists of 3 to 5 million cycles, and the test frequency is 1 Hz, the duration of one test is 6 to 10 weeks including the periodic wear measurements. Therefore, it is obvious that a wear test machine must have several test stations so that more than one sample can be tested simultaneously. The more samples that can be run at the same time, the better.

Chapter written by Vesa SAIKKO.

Hence the capacity of the test device is of great importance in contemporary wear study programs.

4.2. Categories of test devices

A wear test machine in this field is either a joint simulator in which actual prosthetic components, such as total hip and knee prostheses are used as specimens, or a so-called screening device, such as a pin-on-disk, in which the test conditions and the shape of the specimens are more or less simplified. However, the purpose is to produce realistic wear mechanisms and wear factors (mm^3/Nm).

For practical reasons, the number of test stations in joint simulators is usually limited to 12. A common practice is to test 4 samples simultaneously, so the *n* value is 3. From the point of view of comparisons, this is a very low value of course, but it has been justified by the fact that testing is expensive, laborious and time-consuming. Even the purchase price of a 12-station hip joint simulator can be hundreds of thousands of Euros.

4.3. CTPOD principle

In wear screening, however, there is currently little excuse for a low number of test stations because it has been shown that it is possible to build a valid, structurally simple 100-station device in which the downtime for periodic wear measurement can be as low as 8 hours with one operator [SAI 05, SAI 10]. This device is a state-of-the-art hip wear simulator called the SuperCTPOD (Figure 4.1), designed to reproduce the clinical wear mechanisms of total hip prostheses.

CTPOD is an abbreviation for circular translation pin-on-disk. Circular translation means that the pin slides along a circular track of 10 mm diameter against the disk, the sliding being pure x-y-translation without any rotations. This makes the principle fundamentally different from the common uniaxial tribometer in which the disk rotates and the direction of sliding relative to the pin does not change. This has been found to lead to wear mechanisms and wear rates that are not in agreement with clinical findings. The same holds true for linear reciprocating motion [SAI 99, SAI 04, WAN 96, WAN 01].

In the CTPOD, the direction of sliding changes continually relative to the pins, which together with serum-based lubricant were shown to be the key factors in the reproduction of clinically valid wear mechanisms, manifested primarily as (a) burnished UHMWPE (ultra-high molecular weight polyethylene) wear surface (Figure 4.2), (b) wear particles mainly in the 0.1 µm to 1.0 µm size range (Figure 4.3), (c) wear factors of the order of 10^{-6} mm^3/Nm with conventional gamma-sterilized UHMWPE against polished CoCr, and (d) an undamaged CoCr counterface [ATK 85a, ATK 85b, CAM 95, HAL 96, HIR 96, KOB 97, MAB 02, McK 95, SCO 01].

The circular translation is actually a polishing motion that can be implemented with a crank and an x-y-table. This makes it possible to have many disks on a motion plate of a reasonable area. In the original 100-station SuperCTPOD design, the table area occupied by the machine is 384 mm × 339 mm only. The disk diameter is 28 mm and the pin diameter is 9 mm. The distance between the centers of adjacent disks is 34 mm. It has been shown that 1.1 MPa is a sensible choice for the nominal contact pressure. The required load, 70.7 N per station, is generated with 100 pneumatic cylinders. In addition, it has been shown that static load is sufficient for realistic wear simulation, and that room temperature is preferred as the test temperature because of serum degradation problems that get worse with increasing temperature. The SuperCTPOD has a water cooling system for keeping the temperature uniform in all test stations (usually 20°C ± 0.5°C). The circulating cooling water surrounds the 100 test chambers that contain the diluted serum lubricant. The brim of the motion plate forms a basin for the cooling water.

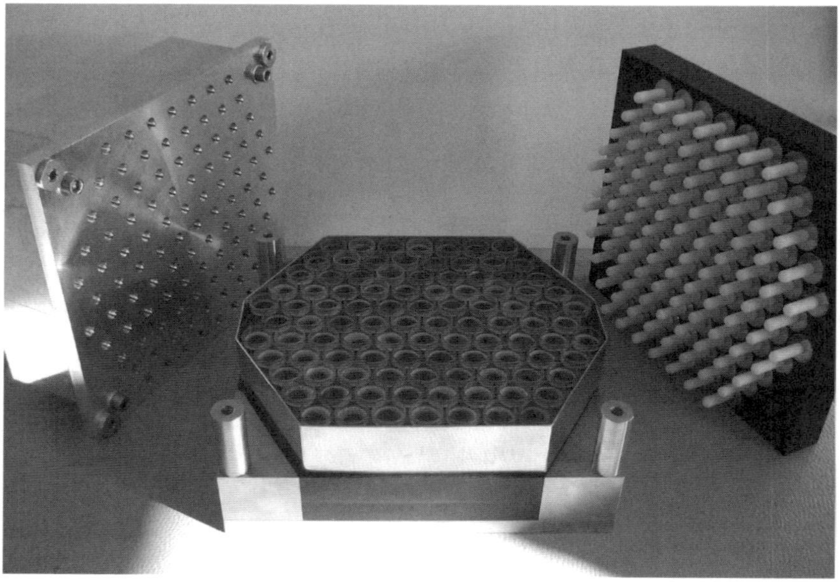

Figure 4.1. *SuperCTPOD hip wear simulator with 100 test stations. The three modules are (from left to right) loading, motion and pin guiding. The modules are ready for assembly and the start of the test. The lubricant chambers on the motion module are filled with serum-based lubricant. The test disk forms the "bottom" of the chamber. The rows are arranged in the 7-8-9-10-11-10-11-10-9-8-7 fashion to allow space for the corner pillars, yet the size of the device (384 mm × 339 mm) is minimized. The small gap between chambers is necessary for the flow of cooling water*

Another practical way to produce circular translation is the so-called parallel crank mechanism with a drive crank and two support cranks, which was utilized in the 12-station CTPOD [SAI 99]. This design was the precursor of the SuperCTPOD, and it still has the second largest capacity among the POD devices. Naturally, an x-y-table could also be driven by two servo-motors, in which case the additional benefit is that many different track shapes can easily be implemented.

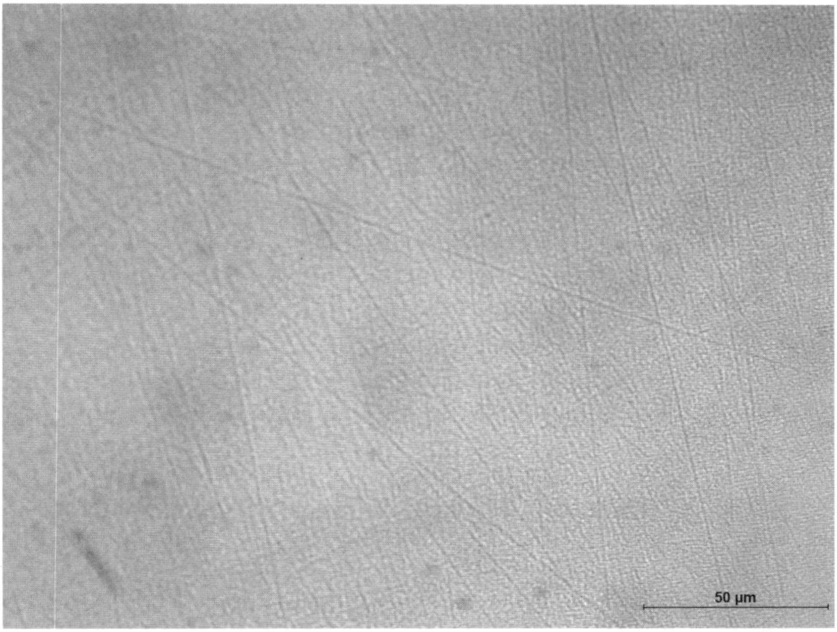

Figure 4.2. *Micrograph from the wear surface of a UHMWPE pin after a SuperCTPOD test. The surface is flat and featureless. Only subtle criss-cross scratches are observed. Visually, the surface is highly polished, which is a result from the combined action of the multidirectional motion and protein-containing lubricant*

4.4. SuperCTPOD test procedure

In the established SuperCTPOD test program, wear is measured once a week gravimetrically, that is, the pins are weighed with a balance having a resolution of 0.01 mg. Since the length of the test is 6 weeks (3 million cycles), there will be 6 points in the graph of wear (mg) versus the number of cycles (million). The wear is typically linear and therefore a straight line is fitted to the 6 data points using the least squares linear regression, the correlation coefficient R^2

values being above 0.99. The slope of the line is the wear rate (mg per one million cycles).

The wear factor is calculated so that the wear rate is divided by the density of UHMWPE (0.94 mg/mm^3), load (70.7 N) and the sliding distance, 31.4 km per one million cycles. For instance, a wear rate of 3.40 mg per one million cycles corresponds to a wear factor of 1.63×10^{-6} mm^3/Nm. The wear factor is a useful quantity as it makes it possible to compare the wear produced in a screening device with the wear produced in a joint simulator, and with clinical wear [ATK 85b, HAL 96]. The laboratory work during the wear measurement stop consists mainly of disassembly, washing, weighing, reassembly, and pipetting of the fresh lubricant. The weekly workload from this procedure is 8 hours for one operator.

Figure 4.3. *Scanning electron micrograph of UHMWPE wear particles on a Nucleopore polycarbonate filter membrane with 0.05 µm pore size. The mean equivalent circle diameter of wear particles produced in this SuperCTPOD test was 0.25 µm. Only separate particles can be used in size distribution analysis. From the agglomerates, individual particles cannot be reliably distinguished*

4.5. SuperCTPOD validation

In the validation study of the test device, the specimens were similar in all test stations (n = 100). The pins were conventional gamma-sterilized GUR 1020 UHMWPE, the disks were polished CoCr, and the lubricant was diluted calf serum with a protein concentration of 21 mg/ml, without additives. The wear mechanisms and wear factors were shown to be realistic, that is, they closely resembled published clinical observations on total hip prostheses [ATK 85a, ATK 85b, CAM 95, HAL 96, HIR 96, KOB 97, MAB 02, McK 95, SCO 01].

Interestingly, it was shown for the first time, by this unparalleled test, that the wear factors were normally distributed. Because of the normality, Student's t-test can be considered appropriate for the comparison of mean wear rates in future tests, even if the n values are much lower so that normality cannot be shown in each case. The S.D. was 5.4% of the mean wear factor that was 1.63×10^{-6} mm^3/Nm. In the calculation of wear rates, the correlation coefficient R^2 of linear regression was 0.9973 on average, which proves not only that the wear was highly linear, but also that the entire test system was consistent.

From the above S.D. value it follows that if the n value is 4, an 8% difference between the sample mean wear rates of two materials results in a p value of 0.045; in other words, the probability that no difference actually exists is only 0.045. It is usually considered that if the p value is below 0.05, we can be reasonably confident that there is a difference in the behavior between the two materials. If it was considered sufficient to distinguish 8% differences, 25 materials, or conditions, could be tested simultaneously.

Materials and conditions to be compared may include other types of polymers and their numerous manufacturing parameters, pin geometry, contact pressure, type of counterface (metallic, ceramic, coating), counterface roughness, micro-patterned counterface, hard-on-hard couples, temperature, and lubricant. The comparison of lubricants would be needed, for instance, in a study aiming to develop a synthetic, stable, inexpensive alternative to serum to be used as a lubricant in wear tests. The inevitable degradation of serum is a true problem in contemporary wear testing activities.

4.6. Further SuperCTPOD studies

In the next phase, 38 different types of polyethylene were compared, the n values varying from 4 to 6 [SAI 10]. In two consecutive tests, 200 pins were worn for 3 million cycles. The test conditions were the same as in the validation study. The mean wear rates of the samples varied from 0.52 to 77.1 mg per one million cycles, and the S.D. values from 1.0 to 8.5% of the mean, being 4.2% on average. The average R^2 value was 0.9978. Two pins showed a remarkable R^2 value of 1.0000.

When sample means were compared in pairs using the t-test, it was found that a difference below 5% resulted in a p value below 0.05 in 21 comparisons. The lowest of such differences was 3.2% ($p = 0.040$), in which case the standard deviations were 2.3 and 2.7% of the means (Figure 4.4). In this comparison, the individual wear rates of the pins were 7.93, 8.28, 8.31, 8.36 and 8.40 mg per one million cycles with material A, and 8.24, 8.41, 8.46, 8.67 and 8.83 mg per one million cycles with material B.

When the difference between two other materials was 4.9%, the largest difference among the above-mentioned 21 comparisons, the p value was as low as 0.001, the S.D.s being

2.0 and 1.0% of the means. The latter S.D. was the lowest obtained in this study. In general, if the S.D. is 4% of the mean and the *n* value is 5, a difference of 5% between the sample mean wear rates results in a *p* value of 0.045. The value of the test statistic in this example being 1.93, the probability that a value of this magnitude or larger is observed if there in fact is no difference between the two materials, is only 0.045.

4.7. Summary

An efficient wear test system has a large capacity and it produces a low S.D. of wear rate. The latter makes it possible to use fairly low *n* values, from 4 to 6, and still smaller differences in wear between materials and conditions result in *p* values below 0.05. Consequently, the number of samples that can be run simultaneously is as high as possible, which together with the large capacity saves valuable time in the test program.

With the SuperCTPOD for instance, 20 to 25 samples can be tested simultaneously. Valid wear simulation naturally is a prerequisite. Without close resemblance to real clinical wear, the testing is pointless, no matter how low the S.D. is and how abundant the data are.

It may be argued that a difference of 5% in laboratory wear rates is so low that it has little clinical relevance considering the numerous design, surgery and patient-specific factors that affect the clinical wear. However, if a wear test method has a tendency to produce high S.D. values and therefore the *p* values are low only with large differences, and if only by increasing *n* values, *p* values below 0.05 are obtained with small differences, the method in question cannot be called state of the art, but the test control obviously needs to be improved.

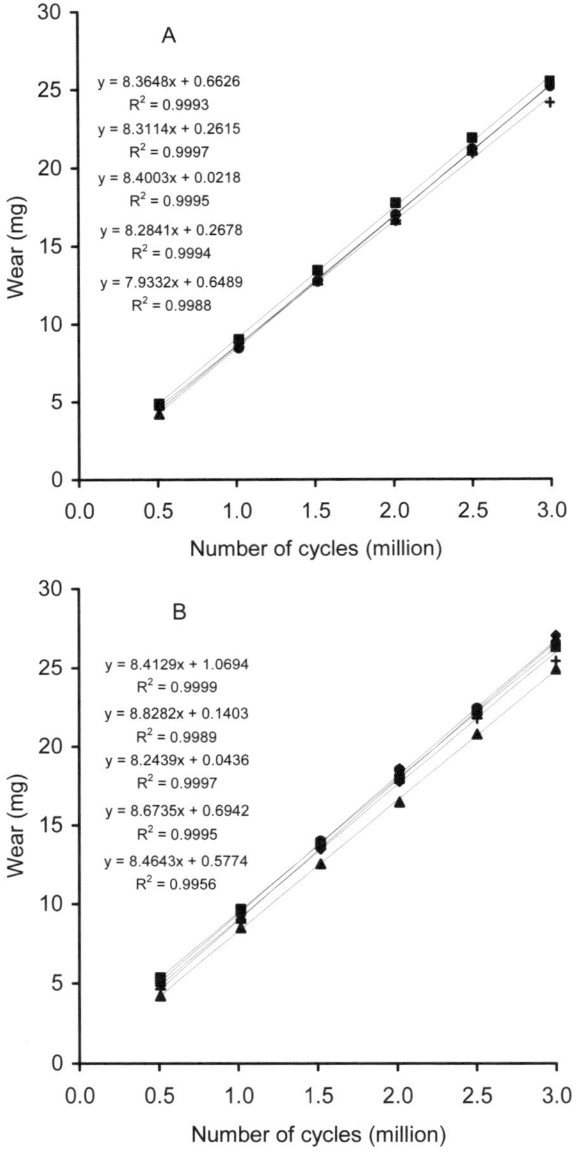

Figure 4.4. *Wear of two samples (n = 5), representing two different types of UHMWPE, in a 6-week SuperCTPOD test with 6 wear measurement points. Note the linearity and uniformity of wear. The difference between the sample mean wear rates is only 3.2%, yet the p value is 0.040 due to low S.D. values that are 2.3 and 2.7% of the mean wear rates*

4.8. Concluding remarks

The emphasis of this chapter is on the capacity of the test device, standard deviation of wear rates produced, comparison of mean wear rates, validity of the wear simulation, and test conditions required for realistic wear. An extensive presentation of statistical techniques in biotribological wear testing can be found in [DUM 81]. In general, it appears that the importance of capacity and S.D. in wear studies of orthopedic biomaterials is eventually being recognized. In the past, the available data was fairly scarce due to the lack of testing capacity. This was considered a normal status to which the decision regarding the introduction of a new material had to adjust.

Biotribology research has now entered a new era as capacity is no longer a bottleneck in wear testing. The commercial SuperCTPOD simulator, TE 87 (Phoenix Tribology Ltd) is available in 50-station and 100-station versions. It has the additional feature of variable slide track shape. The track can be adjusted from a circle, through ellipses with different aspect ratios, to a straight line, in which case the machine is a linear reciprocator. The current users of the TE 87 include universities, contract testing services, manufacturers of orthopedic biomaterials, and manufacturers of orthopedic implants.

4.9. Acknowledgments

The author would like to thank the Academy of Finland (project Biomimetic water lubrication) for financial support, and DSM Biomedical Materials B.V., the Netherlands, for the permission to use material from a joint research project.

4.10. Bibliography

[ATK 85a] ATKINSON J.R., DOWSON D., ISAAC G.H., WROBLEWSKI B.M., "Laboratory wear tests and clinical observations of the penetration of femoral heads into acetabular cups in total replacement hip joints II: A microscopical study of the surfaces of Charnley polyethylene acetabular sockets", *Wear*, vol. 104 no. 3, p. 217-224, 1985.

[ATK 85b] ATKINSON J.R., DOWSON D., ISAAC G.H., WROBLEWSKI B.M., "Laboratory wear tests and clinical observations of the penetration of femoral heads into acetabular cups in total replacement hip joints III: The measurement of internal volume changes in explanted Charnley sockets after 2 – 16 years in vivo and the determination of wear factors", *Wear*, vol. 104 no. 3, p. 225-244, 1985.

[CAM 95] CAMPBELL P., MA S., YEOM B., MCKELLOP H.A., SCHMALZRIED T.P., AMSTUTZ H.C., "Isolation of predominantly submicron-sized UHMWPE wear particles from periprosthetic tissues", *Journal of Biomedical Materials Research*, vol. 29 no. 1, p. 127-131, 1995.

[DUM 81] DUMBLETON J.H., "The role of statistics in wear data evaluation" in *Tribology of Natural and Artificial Joints*, Amsterdam, The Netherlands, Elsevier Scientific Publishing Company, 1981.

[HAL 96] HALL R.M., UNSWORTH A., SINEY P., WROBLEWSKI B.M., "Wear in retrieved Charnley acetabular sockets", *Journal of Engineering in Medicine*, vol. 210 no. H3, p. 197-207, 1996.

[HIR 96] HIRAKAWA K., BAUER T.W., STULBERG B.N., WILDE A.H., "Comparison and quantitation of wear debris of failed total hip and total knee arthroplasty", *Journal of Biomedical Materials Research*, vol. 31 no. 2, p. 257-263, 1996.

[KOB 97] KOBAYASHI A., BONFIELD W., KADOYA Y., YAMAC T., FREEMAN M.A.R., SCOTT G., REVELL P.A., "The size and shape of particulate polyethylene wear debris in total joint replacements", *Journal of Engineering in Medicine*, vol. 211 no. H1, p. 11-15, 1997.

[MAB 02] MABREY J.D., AFSAR-KESHMIRI A., ENGH G.A., SYCHTERZ C.J., WIRTH M.A., ROCKWOOD C.A., AGRAWAL C.M., "Standardized analysis of UHMWPE wear particles from failed total joint arthroplasties", *Journal of Biomedical Materials Research (Applied Biomaterials)*, vol. 63 no. 5, p. 475-483, 2002.

[McK 95] MCKELLOP H.A., CAMPBELL P., PARK, S.-H., SCHMALZRIED T.P., GRIGORIS P., AMSTUTZ, H.C., SARMIENTO A., "The origin of submicron polyethylene wear debris in total hip arthroplasty", *Clinical Orthopaedics and Related Research*, vol. 311, p. 3-20, 1995.

[SAI 99] SAIKKO V., AHLROOS T., "Type of motion and lubricant in wear simulation of polyethylene acetabular cup", *Journal of Engineering in Medicine*, vol. 213 no. H4, p. 301-310, 1999.

[SAI 04] SAIKKO V., CALONIUS O., KERÄNEN J., "Effect of slide track shape on the wear of ultra-high molecular weight polyethylene in a pin-on-disk wear simulation of total hip prosthesis", *Journal of Biomedical Materials Research (Applied Biomaterials)*, vol. 69B no. 2, p. 141-148, 2004.

[SAI 05] SAIKKO V., "A hip wear simulator with 100 test stations", *Journal of Engineering in Medicine*, vol. 219 no. H5, p. 309-318, 2005.

[SAI 10] SAIKKO V., "Performance analysis of an orthopedic biomaterial 100-station wear test system", *Journal of Mechanical Engineering Science*, vol. 224 no. C3, p. 697-701, 2010.

[SCO 01] SCOTT M., WIDDING K., JANI S., "Do current wear particle isolation procedures underestimate the number of particles generated by prosthetic bearing components?", *Wear*, vol. 251 no. 1-12, p. 1213-1217, 2001.

[WAN 96] WANG A., STARK C., DUMBLETON J.H., "Mechanistic and morphological origins of ultra-high molecular weight polyethylene wear debris in total joint replacemant prostheses", *Journal of Engineering in Medicine*, vol. 210 no. H3, p. 141-155, 1996.

[WAN 01] WANG A., "A unified theory of wear for ultra-high molecular weight polyethylene in multi-directional sliding", *Wear*, vol. 248 no. 1-2, p. 38-47, 2001.

Chapter 5

Biotribology of Titanium Alloys

This chapter introduces the application of titanium alloys in biomedical fields. Several different surface modifications of titanium alloys including ion implantation, carburization, and micro arc dioxide are presented in this chapter. It is mainly focused on the biotribological performance, mechanical properties and biocompatibility of the modified titanium alloys.

5.1. Introduction

5.1.1. *History of titanium alloys*

Titanium is known as a transition metal denoted on the periodic table of elements by the symbol Ti. Titanium was discovered in 1791 by the Reverend William Gregor, an English pastor. The element was rediscovered a few years later in the ore rutile by M.H. Klaproth, a German chemist. Klaproth named the element titanium after the mythological giants, the Titans. As titanium reacts easily with non-metallic elements such as oxygen, nitrogen, hydrogen and

---

Chapter written by Yong LUO.

carbon, it is very difficult to obtain pure titanium. The first significant isolation of nearly pure titanium was accomplished in 1875 by Kirillov in Russia. Isolation of the pure metal was not demonstrated until 1910 when Matthew Hunter and his associates caused titanium tetrachloride to react with sodium in a heated steel bomb. This process produced individual pieces of pure titanium. These demonstrations prompted William Kroll to begin experimenting with different methods for efficiently isolating titanium. These early experiments led to the development of a process for isolating titanium by reduction with magnesium in 1937.

This process, now called the Kroll process, is still the primary process for producing titanium. The first products made from titanium were introduced around the 1940s and included wires, sheets, and rods. While Kroll's work demonstrated a method for titanium production on a laboratory scale, it took nearly a decade longer before it could be adapted for large-scale production. In 1948, DuPont opened the first large scale manufacturing operation.

This large scale manufacturing method allowed for the use of titanium as a structural material. In the 1950s, it was used primarily by the aerospace industry in the construction of aircraft. Since titanium was superior to steel for many applications, the industry grew rapidly. Numerous new applications and industries for titanium and its alloys have been discovered over the years. Today, about 80% of titanium is used by the aerospace industry and 20% by non-aerospace industries.

5.1.2. *The properties of titanium alloys*

Titanium is a strong, light metal. It is as strong as steel and twice as strong as aluminum, but is 45% lighter than steel and only 60% heavier than aluminum. It is a lightweight, silver-gray material with an atomic number of

22 and an atomic weight of 47.90. It has a density of 4,510 kg/m^3, which is somewhere between the densities of aluminum and stainless steel. It has an approximate melting point of 1,667°C and a boiling point of 3,287°C. Titanium has chemically similar behavior to zirconium and silicon, and has excellent corrosion resistance and a high strength to weight ratio.

Titanium is the fourth most abundant metal making up about 0.62% of the Earth's crust. Rarely found in its pure form, titanium typically exists in minerals such as anatase, brookite, ilmenite, leucoxene, perovskite, rutile and sphene. While titanium is relatively abundant, it continues to be expensive because it is difficult to isolate.

Titanium is not easily corroded by sea water and is used in propeller shafts, rigging and other parts of boats that are exposed to sea water. Titanium and titanium alloys are used in airplanes, missiles and rockets where strength, low weight and resistance to high temperatures are important. Since titanium does not react within the human body, it is used to create artificial hips, pins for setting bones and for other biological implants. Unfortunately, the high cost of titanium has limited its widespread use.

5.1.3. *The application of titanium alloys*

Since the introduction of titanium and titanium alloys in the early 1950s, these materials have in a relatively short time become backbone materials for the aerospace, energy and chemical industries.

Thousands of titanium alloys have been developed and these can be grouped into four main categories. Their properties depend on their basic chemical structure and the way they are manipulated during manufacture. Some elements used for making alloys include aluminum, molybdenum, cobalt, zirconium, tin and vanadium. Alpha phase alloys have the lowest strength but are formable and able to be welded. Alpha plus (+) beta alloys have high strength. Near alpha alloys have medium strength but have good creep resistance. Beta phase alloys have the highest strength of any titanium alloys but they also lack ductility.

The applications of titanium and its alloys are numerous. The aerospace industry is the largest user of titanium products. Titanium is useful for this industry because of its high strength to weight ratio and high temperature properties. It is typically used for airplane parts and fasteners. These same properties make titanium useful for the production of gas turbine engines. It is used for parts such as the compressor blades, casings, engine cowlings, and heat shields.

Since titanium has good corrosion resistance, it is an important material for the metal finishing industry. Here it is used for making heat exchanger coils, jigs and linings. Titanium's resistance to chlorine and acid makes it an important material in chemical processing. It is used for the various pumps, valves and heat exchangers on the chemical production line. The oil refining industry employs titanium materials for condenser tubes because of corrosion resistance. This property also makes it useful for equipment used in the desalinization process.

Titanium is used in the production of human implants because it has good compatibility with the human body. One of the most notable recent uses of titanium is in artificial hearts, first implanted in a human in 2001. Other uses of titanium are in hip replacements, pacemakers, defibrillators, and elbow and hip joints.

Finally, titanium materials are used in the production of numerous consumer products. It is used in the manufacture of things such as shoes, jewelry, computers, sporting equipment, watches and sculptures. Titanium dioxide is used as a white pigment in plastic, paper and paint. It is even used as a white food coloring and as a sunscreen in cosmetic products.

5.2. Surface modification of titanium alloys

5.2.1. *Ion implantation*

Ion implantation is a material engineering process by which ions of a material can be implanted into another solid, thereby changing the physical properties of the solid. Ion implantation is used in semiconductor device fabrication and in metal finishing, as well as various applications in materials science research.

The ions introduce both a chemical change in the target, in that they can be a different element from the target or induce a nuclear transmutation, and a structural change – the crystal structure of the target can be damaged or even destroyed by the energetic collision cascades.

Among the ion implantation, carbon, oxygen, and nitrogen are widely used to modify the surface of material, with the improvement of surface hardness, wear resistance, wettability and corrosion resistance.

In this section, the nitrogen ion implantation on titanium alloys was introduced as described in [LUO 09a]. Titanium alloys (Ti6Al4V) were machined in a square shape of 15 mm × 15 mm, and 5 mm in thickness. Before performing implantation, Ti6Al4V samples were polished, and then they were ultrasonically cleaned in acetone for 15 min. The nitrogen ion implantation was made in a multi-function ion implanting system, with pressure of 10^{-5} Pa and accelerating energy of 40 keV, and a nitrogen dose of 2.1×10^{17} ions/cm^2 was applied during ion implantation.

The implanted titanium samples were examined by X-ray photoelectron spectroscopy using a VG Thermo ESCALAB 250 spectrometer fitted with monochromatic Al-Kα X-ray radiation (150 W, hν = 1486.6 eV). Analysis spectra were collected with the pass energy of 550 eV in an increment of 1 eV, but high-resolution spectra were collected at 20 eV pass energy in an increment of 0.05 eV.

Element measurements were carried out at 20 depth levels, which were generated by intermittent sputtering for 100 s with argon ions at 3 keV and current intensity of 5 μA. The base pressure in the chamber was 3.1×10^{-8} Pa, and the diameter of the analysis spot was 500 μm. Sputtering rate was approximately in the order of 10 nm per level on the basis of the titanium-sputtering rate.

The XPS peak positions (electron binding energies (BE) for specific atomic levels) are characteristics of chemical states of titanium alloys. Figure 5.1 shows the typical X-ray photoelectron spectra of Ti6Al4V surfaces implanted with 2.1×10^{17} ions/cm^2 nitrogen dose.

Figure 5.1a provides XPS surveys between binding energies 0 and 550 eV at the surface and 1,000 s sputtering depth, in which a series of N1s, Ti2p, O1s, Al2p, and V2p spectra are observed. According to NIST X-ray photoelectron spectroscopy database, the peak at binding energy of 454.9 eV and 460.7 eV in Figure 5.1b corresponds to $Ti2p_{3/2}$ and $Ti2p_{1/2}$ components, which attributed to the TiN phase. The N1s peak at 397.2 eV in Figure 5.1d also corresponds to the formation of a TiN layer. The Ti3p peak at 35 eV and $Ti2p_{3/2}$ peak at 458 eV can be interpreted as a TiO_2 component with the evidence of a O1s peak at 531.3 eV in Figure 5.1c, which results from the sample surface contamination during atmospheric exposure [XIA 98, XIN 98, WAN 02].

In addition, the Ar2p peak at approximately 120 eV present in Figure 5.1a is due to the Ar^+ ion sputter cleaning process. After 1,000 s sputtering, the XPS spectra in Figures 5.1b and c shows apparent reduction of $Ti2p_{3/2}$ peak at 458 eV, and the O1s peak becomes very weak, which suggests that the TiO_2 phase disappears at a certain layer depth. However, the strong peak of N1s at 397.2 eV proves the abundant TiN formation.

The nano-indentation on the Ti6Al4V sample and nitrogen ion implanted samples was done on an *in situ* nano-mechanics testing system (TriboIndenter, Hysitron) with a peak load of 2 mN, which can monitor and record the applied load and displacement of the indenter with a force resolution of 1 nN and displacement resolution of 0.04 nm.

A three-sided pyramid diamond Berkovich indenter with a normal angle of 65.3° was used in nano-indentation tests. A three-stage mode of loading and unloading on indent tip was applied in nano-indentation tests of Ti6Al4V alloys.

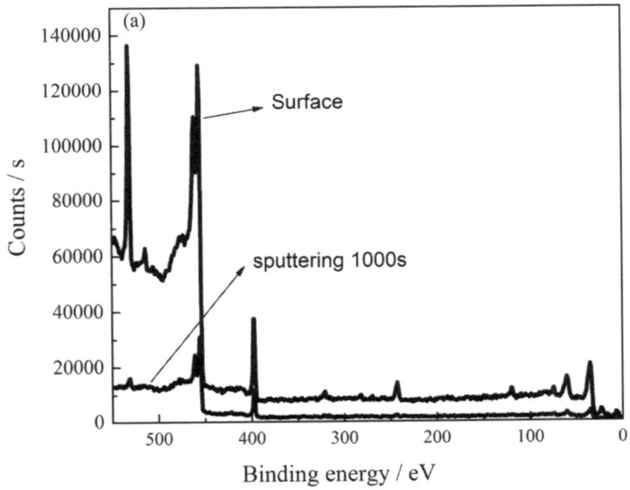

Figure 5.1. *The XPS spectra of nitrogen implanted a Ti6Al4V alloy at the dose of 2.1×10^{17} ions / cm^2 on the surface and 1,000 s sputtering depth regarding (a) the whole spectra*

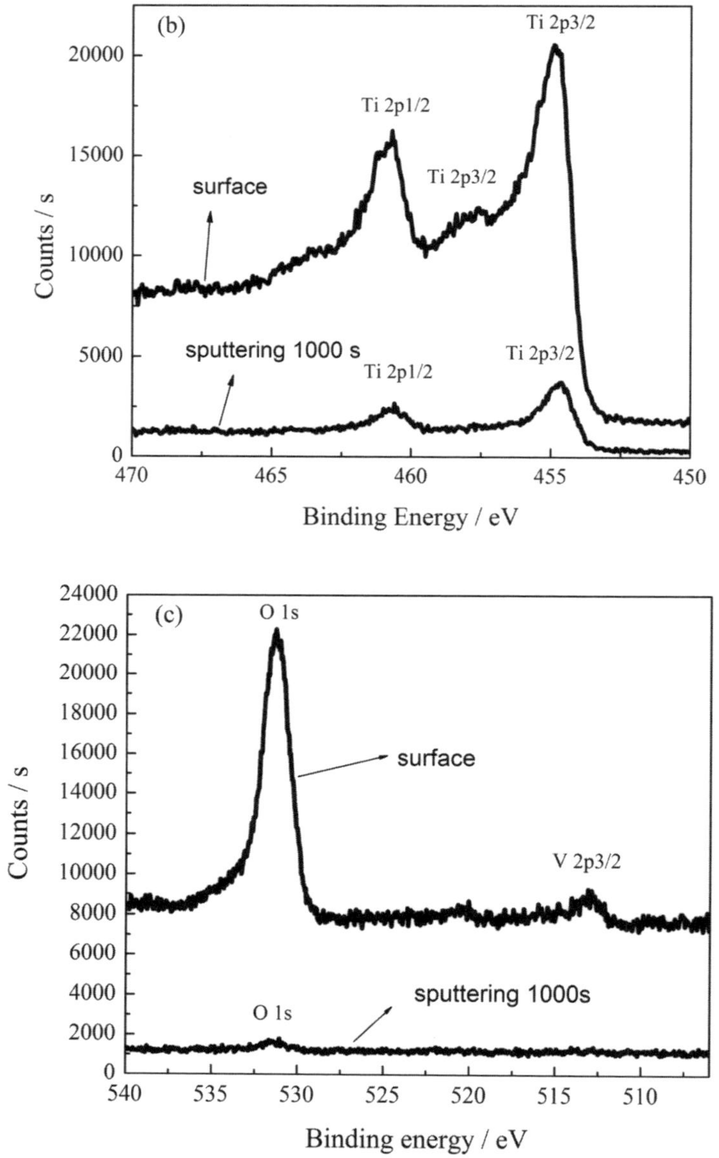

Figure 5.1(continued). *The XPS spectra of nitrogen implanted Ti6Al4V alloy at the dose of* 2.1×10^{17} *ions / cm*^{2} *on the surface and 1000 s sputtering depth regarding (b) the titanium spectrum, and (c) the oxygen spectrum*

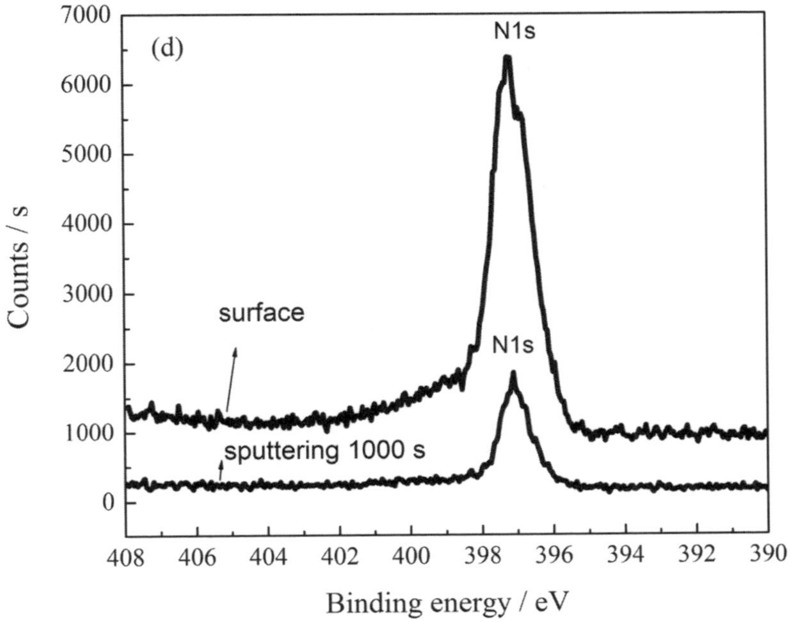

Figure 5.1(continued). *The XPS spectra of nitrogen implanted Ti6Al4V alloy at the dose of 2.1×10*^{17} *ions/cm*^{2} *on the surface and 1,000 s sputtering depth regarding (d) the nitrogen spectrum*

The nanohardness values of nitrogen ion implanted titanium alloys and unimplanted Ti6Al4V samples under the applied loads of 2 mN are shown in Figure 5.2. The nanohardness is determined by load-displacement curves according to Oliver and Pharr methods.

It is shown that the nanohardness value increases from 6.4 GPa of titanium alloy to 7.7 GPa of nitrogen ion implanted titanium alloys with the dose of 2.1×10^{17} ions/cm^2, which results in 20% growth in nanohardness. Such nanohardness enhancement can contribute to higher stiffness of the TiN hard layer formed on N^+ implanted Ti6Al4V substrates.

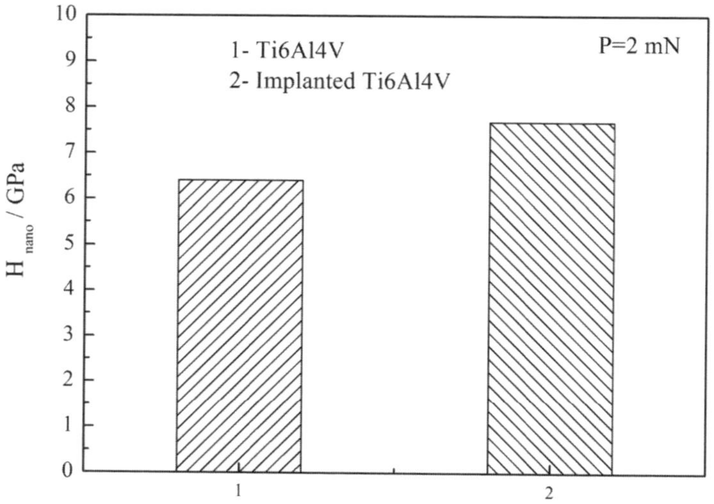

Figure 5.2. *The nanohardness of the nitrogen ion implanted and unimplanted Ti6Al4V alloy*

5.2.2. *Carburization*

Titanium carbide has attracted considerable attention because of its high melting point (3,260°C), low density (4.93 g/cm^3), high hardness (2,800 HV), greater chemical and thermal stability, and excellent outstanding wear resistance [YIN 05]. Titanium carbide is a very important non-oxide ceramic widely used in the fields of wear resistance tools and materials [LEE 03].

In recent years, titanium carbide coatings have been considered as a potential biomedical material especially in biotribological field. It is believed that TiC coating has an ability to form strong bonds with the titanium substrate which could provide a superior hardness and wear resistance, and it can induce carbon atoms into the metal matrix to enhance the biocompatibility. Therefore, forming a TiC coating on biomaterials especially titanium alloys has attracted extraordinary attention due to their potential application in the biotribological field. Brama *et al.* [BRA 07] produced a titanium carbide layer on titanium substrates by pulsed laser deposition and evaluated the biological response both *in vitro* and *in vivo*. It was found that a mixture of TiC, Ti_2O_3, TiO_2 and TiO formed on the surface. It was concluded that coating with TiC will, in comparison to uncoated titanium, improve implant hardness, biocompatibility through surface stability and osseointegration through improved bone growth.

In this section, a novel sequential carburization is used to obtain a pure titanium carbide layer on a titanium alloy surface [LUO 10]. The synthesis process of TiC micro-porous layer is briefly described as follows. A medical grade titanium alloy (Ti6Al4V) specimens in square shape of 15 mm × 15 mm and 5 mm thickness were polished to Ra < 0.05 μm, then ultrasonically cleaned in ethanol and acetone, in turn, for 30 min. and inserted into a vacuum gas carburizing furnace.

Prior to heating, the reaction chamber was first evacuated to reach a system pressure of 1 Pa using a vacuum pump. After that, the chamber was quickly heated to high temperature and the pressure was kept at 1 Pa. Subsequently, pumping stopped and an ultrahigh-purity C_2H_2 flow was flushed into the chamber. When the pressure reached 2,000 Pa, the C_2H_2 flow was turned off and pumped again to eliminate the H_2 decompounded by C_2H_2. Then the chamber temperature was kept at high temperature for a while to carburize.

This process is called a carburization step. After the carburization step was repeated several times the furnace was cooled to room temperature. This procedure is called the sequential diffusion method and after the carburization, it was observed that the as-synthesized carburized layer with deep-black color and a porous structure had formed.

The XRD pattern (Figure 5.3) from the as-synthesized bulk samples reveals the crystal structure and phase purity of the products. All the diffraction peaks can be indexed to the cubic structure of TiC, with lattice constants of a = 4.328 Å, which is consistent with the standard value for TiC. There is no evidence of hydrogen which is considered to be a significant source leading to brittle fracture.

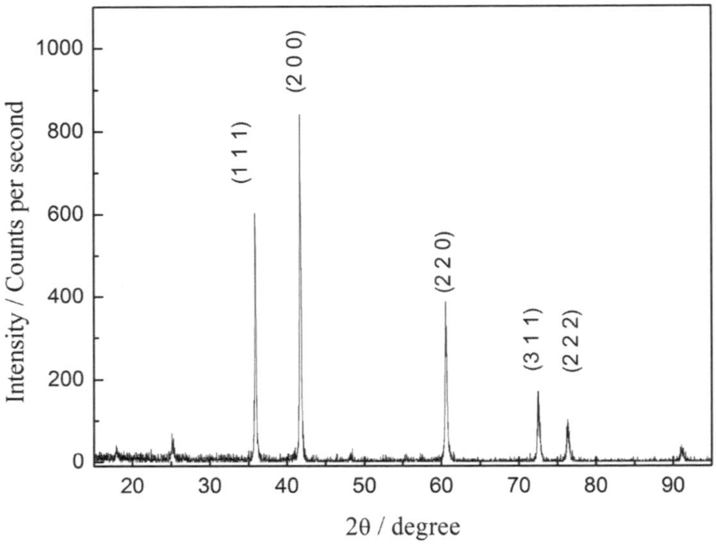

Figure 5.3. *The XRD pattern recorded from the synthesized TiC layer*

The XPS peak positions (electron binding energies (BE) for specific atomic levels) can be used for identification of chemical states. Figure 5.4 shows the typical X-ray photoelectron spectra of the as-synthesized TiC layer at a depth of 40 nm, including a series of Ti2p, C1s, O1s, and Al2p. The peak at binding energy of 451.1 eV and 460.9 eV in Figure 5.4a corresponds to Ti2p3/2 and Ti2p1/2 components, which attributed to the TiC phase [BEN 01, SHI 04]. We note that two strong C1s peaks appear at 281.6 eV and 284.7 eV in Figure 5.4b, which corresponds to the TiC phase and graphite phase. As the C_2H_2 decomposes into active carbon atoms at high temperature, the active carbon atoms will diffuse into the titanium alloy; one of which will react with titanium and form TiC phase, the other will remain in the graphite phase. The Al2p peak at 74.9 eV in Figure 5.4c and the O1s peak at 531.7 eV in Figure 5.4d are both interpreted as Al_2O_3 [DIC 90], which indicate O has a predominance to react with Al rather than Ti.

However, the XRD pattern in Figure 5.3 shows the Al_2O_3 amount is negligible and can be eliminated in a modified model with Ar flow protection in future. All of the results indicate a great quantity of TiC is formed during the carburization and precipitate graphite around the micro-porous TiC.

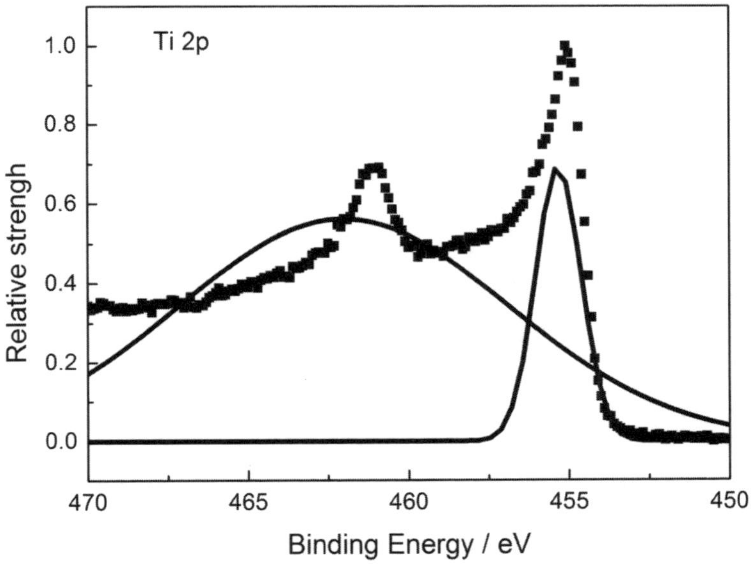

(a)

Figure 5.4. *a) The XPS spectra of the as-synthesized TiC at 40 nm depth*

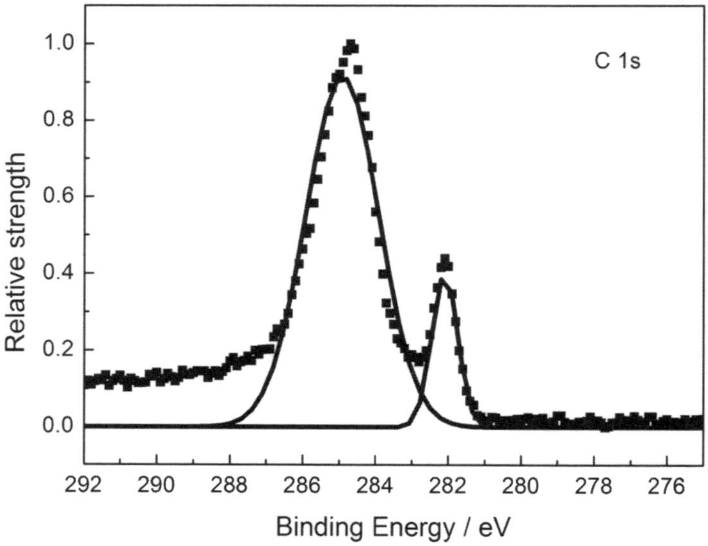

(b)

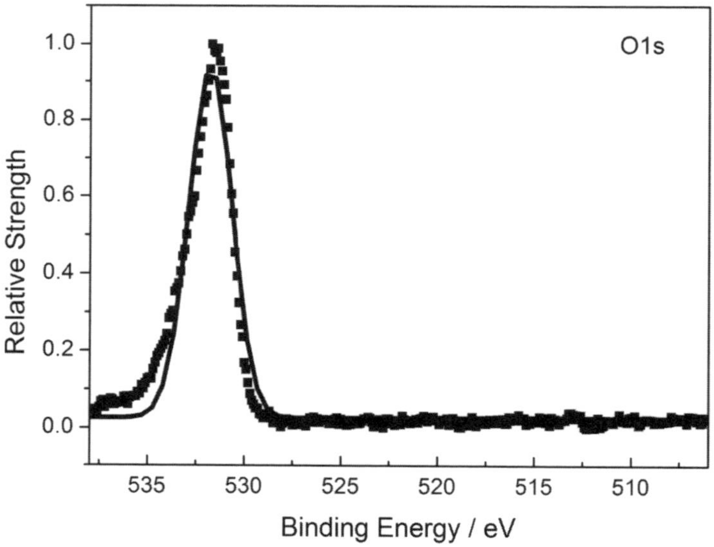

(c)

Figure 5.4. (continued) *b) and c) the XPS spectra of the as-synthesized TiC at 40 nm depth*

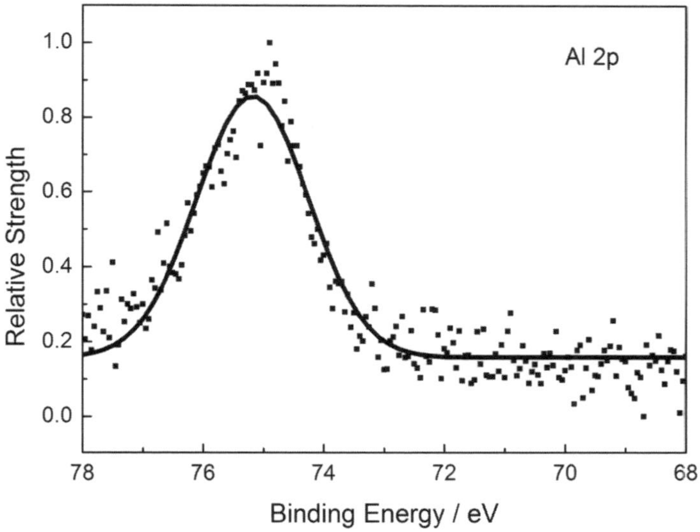

(d)

Figure 5.4(continued). *d) The XPS spectra of the as-synthesized TiC at 40 nm depth*

In order to identify the depth distribution of chemical phase and composition, a depth etching was performed. Each etching time is 170 s which corresponds to a depth of 20 nm and the total depth for the 32 etching times obtains an etching depth of 600 nm. Figure 5.5 presents the depth variety of Ti2p, O1s and C1s, and it is obviously that the bottom curve corresponding to the surface is quite different from the others.

As it is generally accepted that titanium alloy can easily form a compact oxide layer with several nanometers' thickness when it is exposed to air, which can effectively prevent further oxidation and improve the biocompatibility. Therefore, the Ti2p peak at 459.0 eV and the O1s peak at 530.0 eV are interpreted as TiO_2, which is due to the surface contamination. From the C1s spectra, it is found that the intensity of the peak at 281.6 eV becomes stronger with the depth, which suggests an increase of the TiC component. At the same time, the intensity of C1s peak at 284.7 eV falls quickly, indicating a great decrease of graphite with depth.

The shift of O1s peak also provides strong evidence that the oxygen will not react with titanium during the carburization, and it doesn't need expensive ultrahigh vacuum facility to implement the carburization which will save numerous costs for producing TiC.

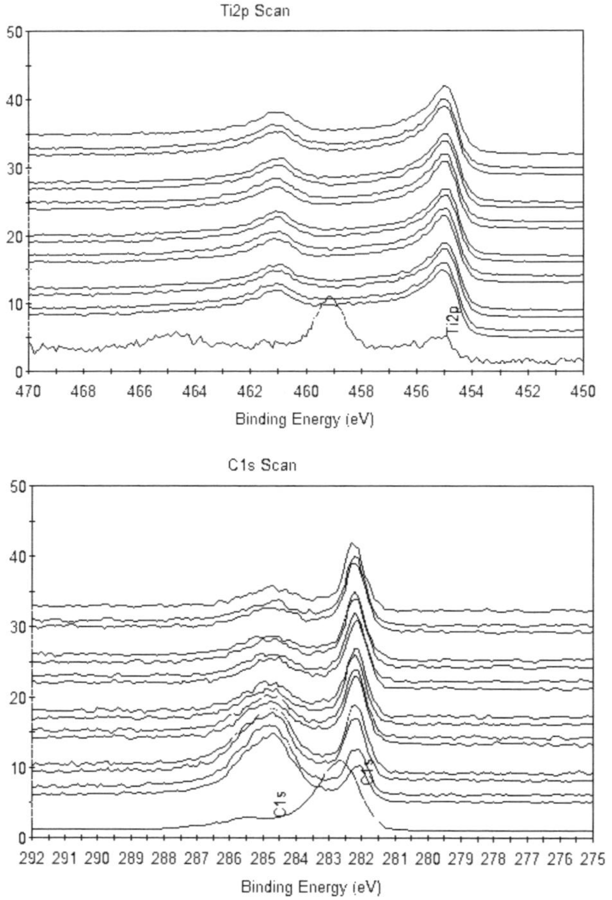

Figure 5.5. *The spectra of Ti2p, O1s and C1s with depth variation: the bottom curve represents the surface, and the depth between two curves is 20 nm*

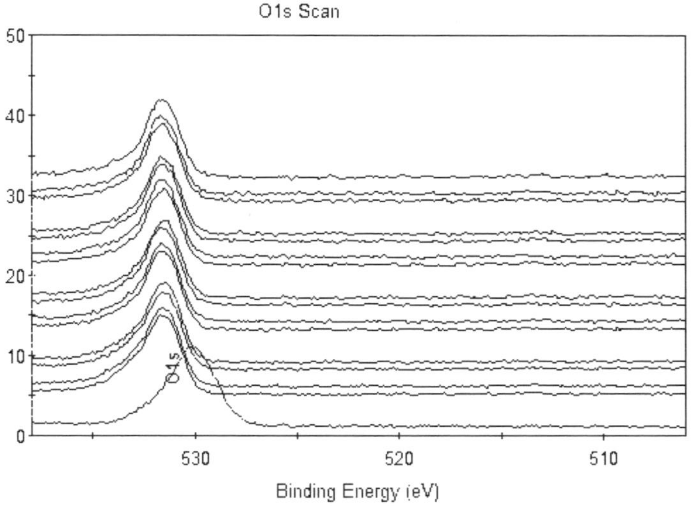

Figure 5.5(continued). *The spectra of Ti2p, O1s and C1s with depth variation: the bottom curve represents the surface, and the depth between two curves is 20 nm*

The atomic percentages of different elements detected by GDMS in variation with etching depth are plotted in Figure 5.6. It is clear that the carbon atomic concentration first declines rapidly due to the decrease of graphite, and then keeps a relatively steady value of 22%, even though a depth of 90 µm supports formation of a large quantity of pure TiC corresponding to XRD results, see Figure 5.3.

The most exciting and significant thing is that the titanium carbide layer is of great thickness, which is quite different from the other surface modification such as PVD and CVD. As a result of the great improvement in thickness of TiC coating, it significantly improves the wear resistance and would lead to a wider application of titanium alloys. Note also that the oxygen atomic concentration decreases rapidly with depth, while the titanium atomic concentration increases quickly. It is concluded that the oxygen doesn't affect the formation of TiC during carburization.

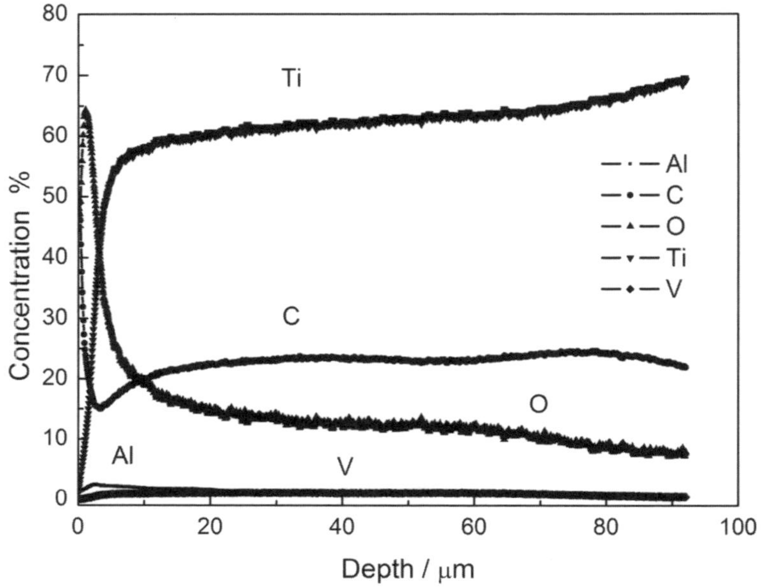

Figure 5.6. *The atomic percentages of different elements detected by GDMS in variation with etching depth*

Figure 5.7 shows the typical SEM image of the as-synthesized products which provides further insight into the micro-porous TiC structure. It was found that the as-synthesized products consist of a large quantity of acicular titanium carbide, with a diameter of around 1 or 2 μm, and forms a uniform micro-porous structure.

It is believed that such a unique structure can deposit lubricant and improve the lubrication, thus reducing friction and decreasing wear, which enables micro-porous TiC to be considered as a potential ultrahigh wear resistance material for biotribological application.

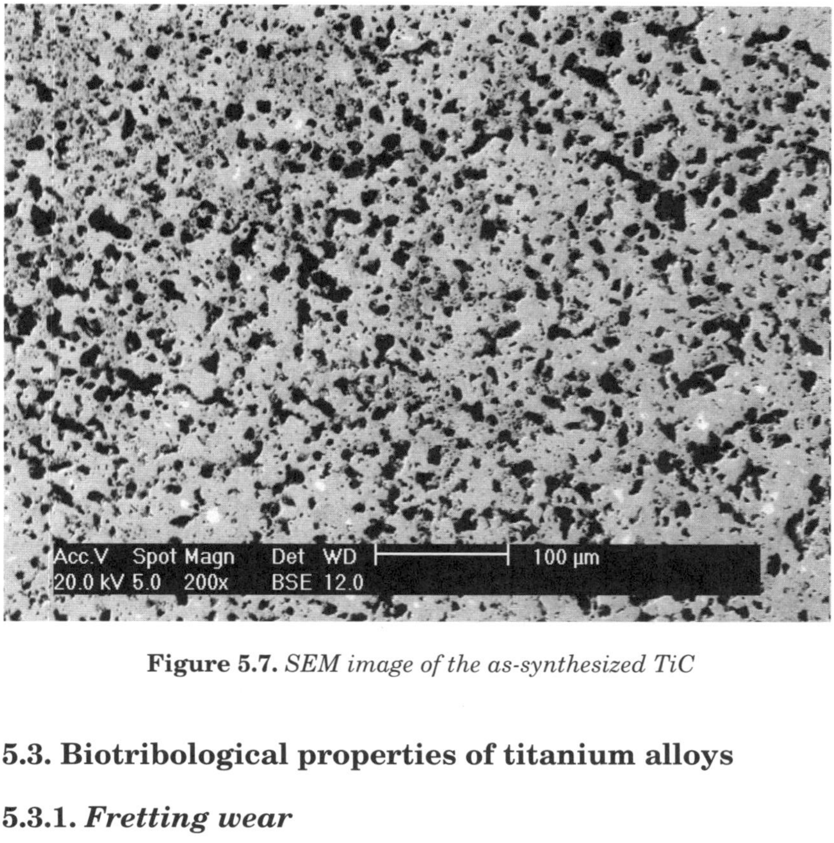

Figure 5.7. *SEM image of the as-synthesized TiC*

5.3. Biotribological properties of titanium alloys

5.3.1. *Fretting wear*

Titanium alloys are generally widely applied for dental and non-cemented orthopedic implants because of their similar modulus to human bone, superior biocompatibility and corrosion resistance.

Recently, there are an increasing number of bio-devices being made of titanium alloys. However, small amplitude oscillatory sliding motion on the titanium alloy implants will result in crack initiation, wear and adhesion of these implants – which are the main failure of titanium alloy implants. It was reported that about 6% of the total hip prostheses will be replaced after 9 years due to aseptic loosening. One of the main causes of the aseptic loosening may be attributed to fretting corrosion on the interfaces of titanium stems and the bone cement.

The fretting wear of nitrogen ion implanted titanium alloys was performed on a UMT tester developed at CETR (Campbell, California) with ball-on-flat fretting style in bovine serum lubrication and dry friction condition, respectively. This tester is able to measure tribological parameters including friction force, *in situ* wear depth, contact acoustic emission, etc.

In order to focus the fretting wear behavior of titanium alloys and the nitrogen implanted samples, an Si_3N_4 ball was chosen for the wear testing counterparts because of its super wear resistance compared to titanium alloy. The schematic of contact between the Si_3N_4 ball and the titanium alloy plate is shown in Figure 5.8. The titanium alloy plate was fixed to the specimen holder in the Y direction, and the Si_3N_4 ball with 4 mm diameter is mounted on a holder connected to a 3D force sensor. New ball specimens were used for each test.

According to our previous experience, it is hard to observe a wear scar for both the titanium alloys and implanted alloys with a normal stress of the joint for a long time. In order to produce a visible scar to compare the fretting behaviors of the samples, a larger normal load of 49.6 N, equivalent stress of 2.68 GPa was applied to the fretting experiment to accelerate the fretting failure of the samples. A reciprocating system based on a motor was designed to induce fretting oscillation of desired amplitude. The Si_3N_4 ball oscillated in a linear velocity of 8 mm/s and reciprocating amplitude of 100 μm. The fretting frequency was approximately 40 Hz. The tangential force and the coefficient of friction were digitally recorded throughout the 25 hour test with a load transducer to explain the contact behavior during fretting.

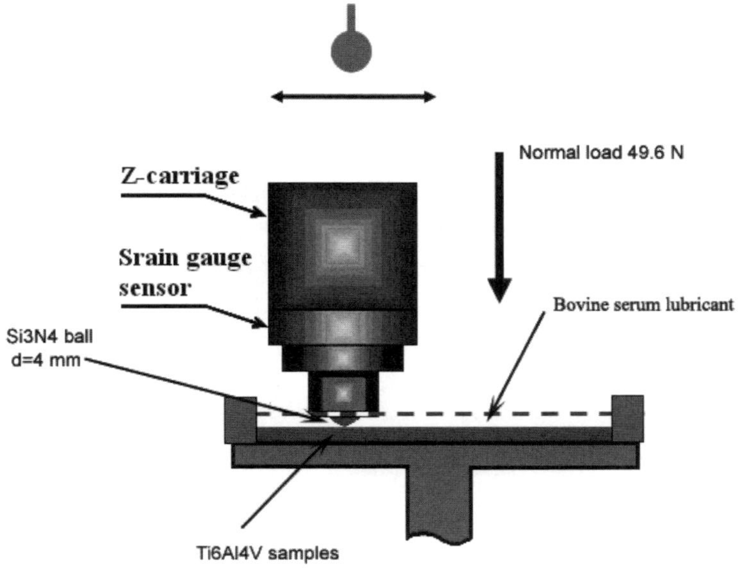

Figure 5.8. *Schematic sketch of fretting wear machine*

Bovine serum was diluted to the concentration of 25%. The temperature was maintained at 37°C during the fretting process to simulate the physiological medium around the contact area of the prostheses within the body.

Figure 5.9 shows the variation of the coefficients of friction of the titanium alloy specimens and nitrogen ion implanted specimens in bovine serum lubrication and dry friction conditions. It is observed that the bovine serum unimplanted Ti6Al4V fretting, which results in the lowest values about 0.056 compared to the coefficients lubrication has the effect of reducing the coefficients of friction by around 0.065 for the unimplanted Ti6Al4V fretting in dry friction condition. For the nitrogen ion implanted Ti6Al4V specimens, it was found that bovine serum lubrication increases the coefficients of friction by about 10% in mean values, which is the opposite of the effect on unimplanted Ti6Al4V specimens. Such results suggest that the nitrogen ion implanted on Ti6Al4V is possibly not beneficial for its application in biological environmental fretting if the lower friction force is stressed. The higher friction force of nitrogen ion implanted Ti6Al4V under bovine serum lubrication is possibly due to protein deposition on fretting surfaces, because nitrogen ion implantation increases the wettability of Ti6Al4V to biological solutions, which provides more adhesion of protein on the rubbing surfaces.

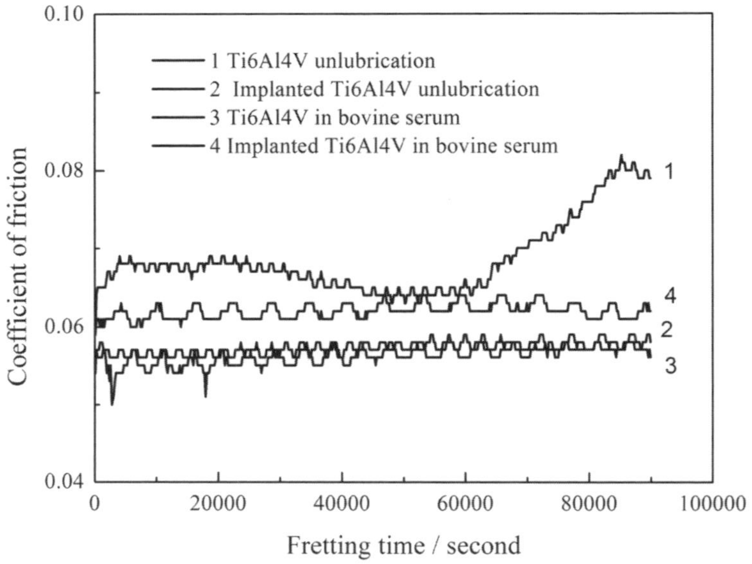

Figure 5.9. *Friction coefficient curves for fretting wear test conducted in dry friction and bovine serum for unmodified and nitrogen ion implanted titanium alloys*

The coefficients of friction of unimplanted Ti6Al4V fretting in dry friction conditions increase at late fretting time, which is quite different from nitrogen ion implanted samples fretting. Figure 5.10 shows the line distribution of oxygen element across the fretting scar of Ti6Al4V alloy in dry friction condition. The strong oxygen element peaks on the fretting area indicate the existence of titanium oxides. It suggests that the titanium oxides are formed on the fretting area due to the frictional heat accumulation during the fretting wear. Thus, the coefficients of friction for Ti6Al4V alloy fretting in dry friction conditions increase at a late phase of fretting testing. However, the bovine serum lubrication can prevent the exposure of the fretting scar to air, which decreases the oxide rate of the titanium and then maintain the steady values of coefficients of friction in fretting process. In addition to the lubrication effect, the nitrogen ion implantation resulting in surface hardness enhancement will reduce the adhesion force, which leads to lower values of friction coefficients [VAD 06].

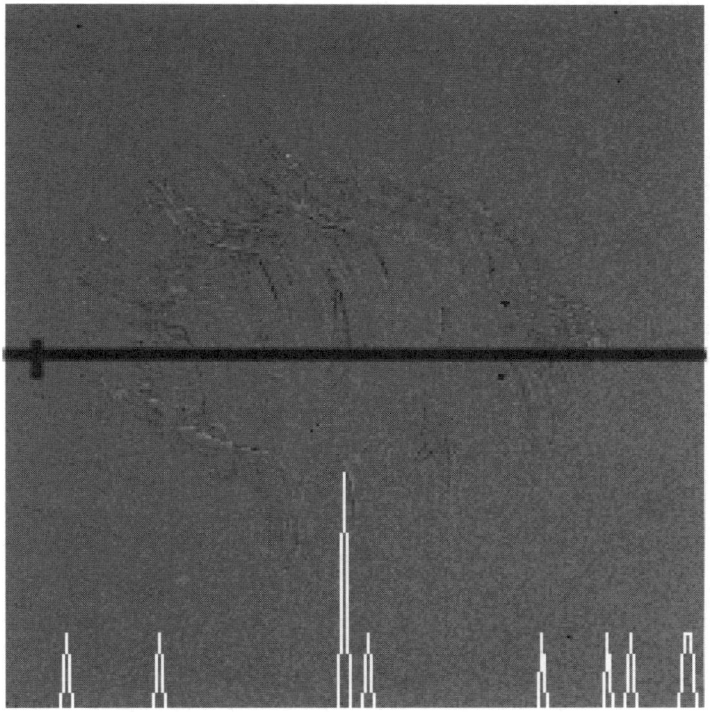

Figure 5.10. *The line distribution of oxygen element on the fretting worn scar of Ti6Al4V alloy in dry friction*

The variation of tangential load in the fretting stroke of Ti6Al4V and nitrogen ion implanted Ti6Al4V in response to reciprocating displacement is shown in Figure 5.11. The elliptical non-symmetrical fretting loop indicates a gross slip mechanism. The envelope area of each fretting loop corresponds to the dissipated energy produced during the fretting circle, which is a very useful parameter to describe the fretting wear mechanism [PAU 05]. We found that the fretting loop of Ti6Al4V in dry friction has the largest envelope area, which reveals a severe adhesion mechanism. The smaller envelope area for the nitrogen ion implanted Ti6Al4V fretting indicates the reduction in adhesion and oxidation wear.

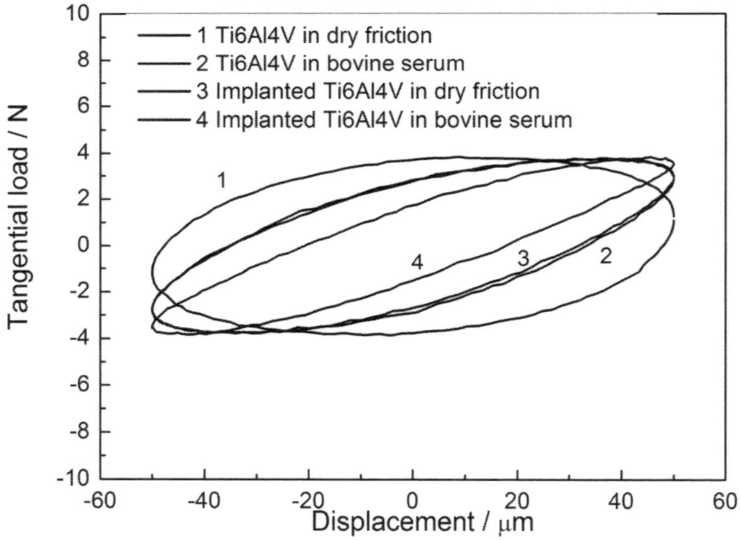

Figure 5.11. *The variation in tangential load of Ti6Al4V and nitrogen ion implanted Ti6Al4V in response to displacement during fretting in dry friction and bovine serum*

SEM micrographs of fretting scar in dry friction and bovine serum lubrication for nitrogen ion implanted Ti6Al4V and the unimplanted Ti6Al4V alloy are shown in Figure

5.12. The fretting scars show evidence of plough and scratch damage on the fretting area. We take the scar diameter, which is defined as the average of the maximum length in horizontal and vertical direction, as a characterization parameter of fretting scar size. The scar diameters for Ti6Al4V in dry friction and bovine serum lubrication are 212 μm and 211 μm, respectively.

For the implanted Ti6Al4V in dry friction and bovine serum lubrication, the fretting scar diameter increases to 235 μm and 237 μm. These results indicate that the bovine serum lubrication has little effect on the variation of fretting scar size for both Ti6Al4V and nitrogen ion implanted Ti6Al4V. However, the nitrogen ion implantation results in the increasing of fretting scar size for Ti6Al4V both in dry friction and bovine serum lubrication.

Figure 5.13 shows the characteristics of the wear damage, which can explain the fretting regime for the Ti6Al4V and implanted Ti6Al4V alloy in dry friction and bovine serum lubrication. Ti6Al4V alloy fretting in dry friction causes a severe plough, scratch and local fracture, which suggests an adhesion and oxidation mechanism for the fretting damage [VAD 07]. In contrast, a slight scratch and local fracture of the nitrogen ion implanted Ti6Al4V alloy fretting in dry friction shows some improvement of wear resistance in dry friction.

At the same time, the crack nucleation as well as crack propagation of the nitrogen ion implanted Ti6Al4V in bovine serum lubrication indicates a fatigue and corrosion mechanism. All of the evidence suggests that nitrogen ion implantation is an effective way to enhance the surface of Ti6Al4V alloy so as to decrease the wear damage both in dry friction and bovine serum lubrication.

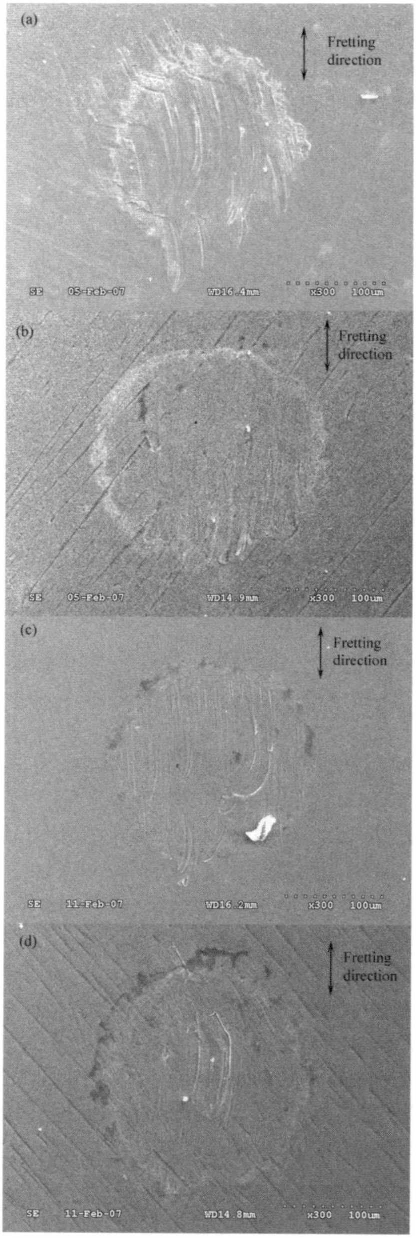

Figure 5.12. *SEM micrographs of fretting scar profiles for (a) Ti6Al4V in dry friction; (b) implanted Ti6Al4V in dry friction; (c) Ti6Al4V in bovine serum lubrication and (d) implanted Ti6Al4V in bovine serum lubrication*

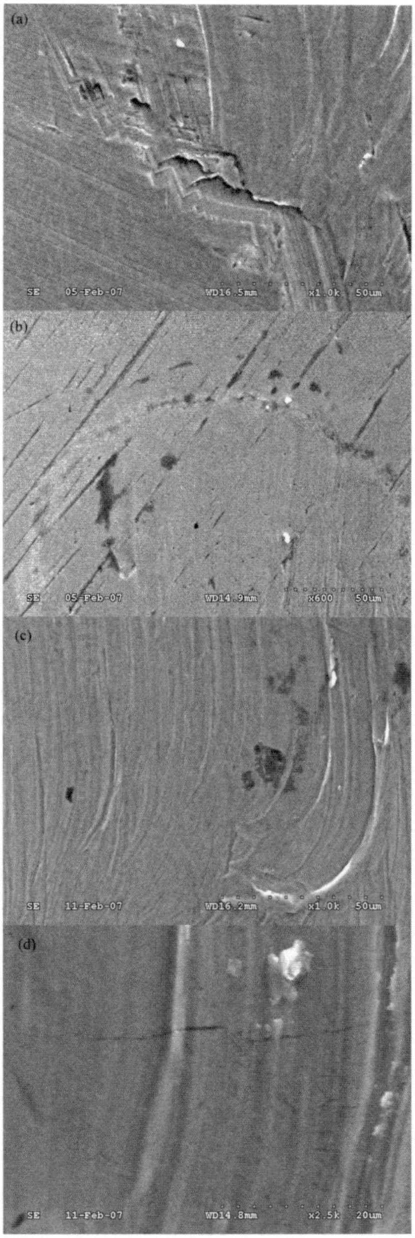

Figure 5.13. *Local SEM micrographs of the fretting scars for (a) Ti6Al4V in dry friction; (b) implanted Ti6Al4V in dry friction; (c) Ti6Al4V in bovine serum lubrication and (d) implanted Ti6Al4V in bovine serum lubrication*

The wear volumes of Ti6Al4V and nitrogen ion implanted Ti6Al4V fretting in dry friction and bovine serum are shown in Figure 5.14. It can be seen that the nitrogen ion implantation decreases the wear volume of Ti6Al4V alloy in dry friction from 6.3×10^4 μm^3 to 3.3×10^4 μm^3, with a reduction of about 90%. Also, nitrogen ion implantation enhances the wear resistance of Ti6Al4V alloy in bovine serum with an increase of 37%. These results indicate that nitrogen ion implantation is effective for fretting wear resistance improvement both in dry friction and bovine serum lubrication.

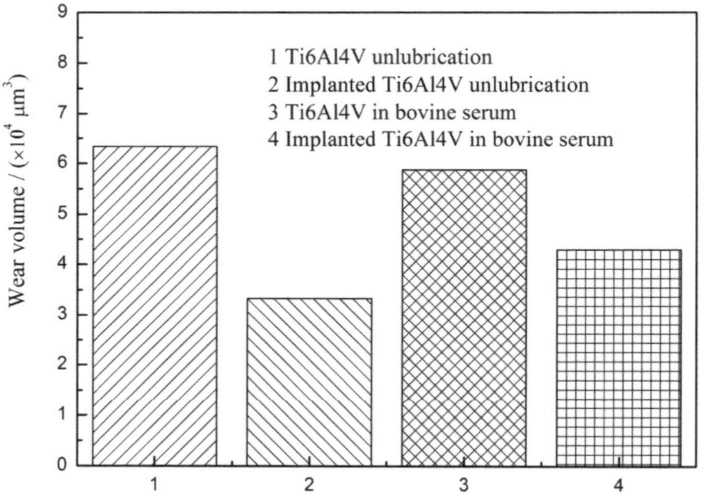

Figure 5.14. *Wear volume for fretting wear tests conducted in dry friction and bovine serum for unmodified and nitrogen ion implanted titanium alloys*

5.3.2. *Sliding wear*

The slide friction and wear behaviors of nitrogen ion implanted and carburized titanium alloys were performed on a UMT tester developed at CETR (Campbell, California) with ball-on-flat configuration in bovine serum lubrication and dry friction conditions, respectively [LUO 09b]. A ZrO_2 ball with a diameter of 5 mm was chosen for the wear testing counterparts due to its super wear resistance, excellent biocompatibility and high toughness. The schematic of contact between the ZrO_2 ball and titanium alloy plate is shown.

The titanium alloy plate was fixed to the specimen holder in the Y direction, and the ZrO_2 ball was mounted on a holder connected to a 3D force sensor. The test for each case repeats three times and new ball specimens were used for each test. A normal load of 1.96 N, equivalent stress of approximately 500 MPa according to Hertz equation, was applied during the 3 hour test with a total journey of 356 m.

It is believed that the contact angle is associated with wettability. A contact angle of less than 90° reveals wettable properties of the material, and lower value of the contact angle corresponds to a better wettability of the material [LUO 09c]. The variation in contact angle of implanted and carburized titanium alloy specimens is shown in Figure 5.15. Note that the contact angle of titanium alloy for distilled water is less than 90°, indicating considerable wettable properties, which makes it easy to understand the superior biocompatibility of medical grade titanium alloy.

It is obvious that the contact angle of titanium alloy decreased after nitrogen ion implantation, from 83° to 80°, with a decrease of 3.6%, indicating a little wettability improvement. At the same time, the contact angle of titanium alloy decreased after carburization, from 83° to 71.5°, with a drop of 14%, significantly enhancing the wettability of titanium. On the other hand, as Tamada concluded, material with a water contact angle of 70° has the best cell adhesion [TAM 93], the contact angle of carburized titanium is much closer to the optimal water contact angle for cell adhesion, which indicates that the carburized titanium alloy is a potential biomedical material.

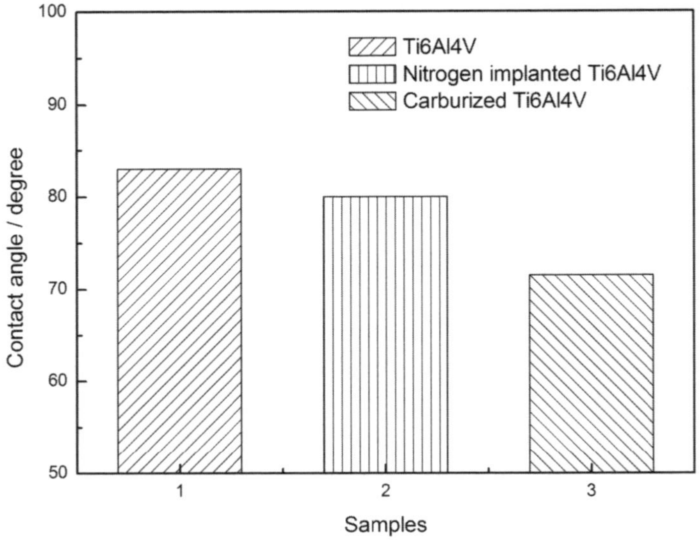

Figure 5.15. *The water contact angle of implanted and carburized titanium alloys*

Figure 5.16 shows the volumetric wear rate of implanted and carburized titanium alloys under bovine serum lubrication. Note that the volumetric wear rate decreased from 12.3×10^{-4} mm^3/m to 5.6×10^{-4} mm^3/m after nitrogen ion implantation, with a decrease of 54.5%, indicating a great improvement of wear resistance. Similarly, the volumetric wear rate decreased from 12.3×10^{-4} mm^3/m to 3.5×10^{-4} mm^3/m after carburization, with a decrease of 71.5%, significantly enhancing the surface of wear resistance.

As previously mentioned the nitrogen ion implantation could form a TiN layer with a thickness of no more than several hundred nanometers, the wear resistance of implanted titanium alloy would weaken after the destruction of the thin layer. Thus nitrogen ion implantation is not an effective method of improving a titanium alloy for a long-term service. On the contrary, carburization can provide titanium carbide layers with a thickness that varies from several hundred micrometers to several millimeters, which can ensure a long-term lower wear rate service. The super wear resistance and thick layer make carburized titanium the applicable to artificial joints.

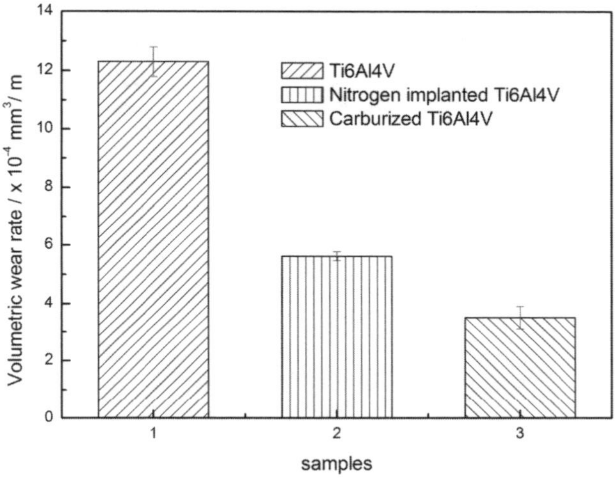

Figure 5.16. *Volumetric wear rate of implanted and carburized titanium alloys under bovine serum lubrication*

SEM micrographs of the worn surfaces for nitrogen ion implanted and carburized titanium alloys are shown in Figure 5.17. Figure 5.17a shows the destruction of the titanium alloy surface, with a severe plough, scratch and fracture, which suggest adhesion and abrasive mechanisms. In contrast, the worn surface of the nitrogen ion implanted titanium alloy shown in Figure 5.17b presents a slight and uniform scratch, which indicates an adhesion mechanism. Figure 5.17c is the worn surface of carburized titanium alloy, in which no obvious destruction or scratch can be found.

The surface of carburized titanium kept the same porous structure and topography after the friction and wear test, and the surface became even smoother. The hard TiC ceramic layer with porous structure was apt to store lubricant and wear debris, which effectively improved the lubrication and avoided abrasive wear, thus significantly strengthened the wear resistance.

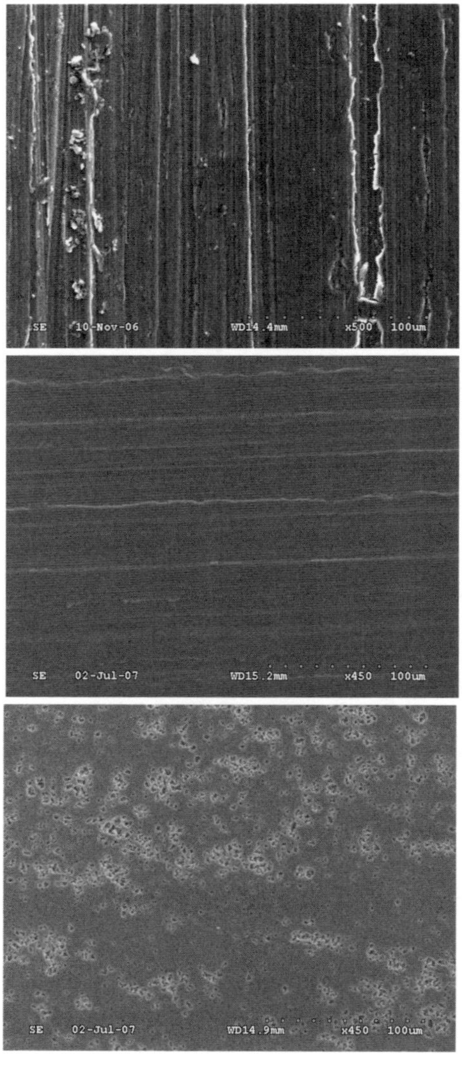

(a) (b) (c)

Figure 5.17. *The worn surface SEM of (a) Ti6Al4V alloy; (b) nitrogen ion implanted titanium alloy; (c) carburized titanium alloy*

Figure 5.18 shows the wear debris distribution of implanted titanium alloy under bovine serum lubrication. Note that the wear debris size of nitrogen ion implanted titanium alloy was between 1 μm and 13 μm, while the wear debris size of titanium alloy was between 2 and 35 μm. The medium radius of wear debris decreased from 6 μm to 4 μm.

Based on the report from Goodman *et al.*, smaller particles could potentially cause more adverse effects on cellular viability, proliferation and function; the specific effects are dependent on particle composition and dose [GOO 06]. This means that nitrogen ion implantation may cause adverse effects on the cellular response with the size decrease of wear debris although the wear resistance has been enhanced.

Note that the laser granularity analyzer could not examine the wear debris distribution of carburized titanium alloy, which may be caused either by very small wear debris size or very low wear debris dose. All of the results suggest that carburized titanium alloy could significantly improve the wear resistance and avoid adverse effects on the cellular response caused by wear debris, indicating a potential biomaterials application in biomedical fields.

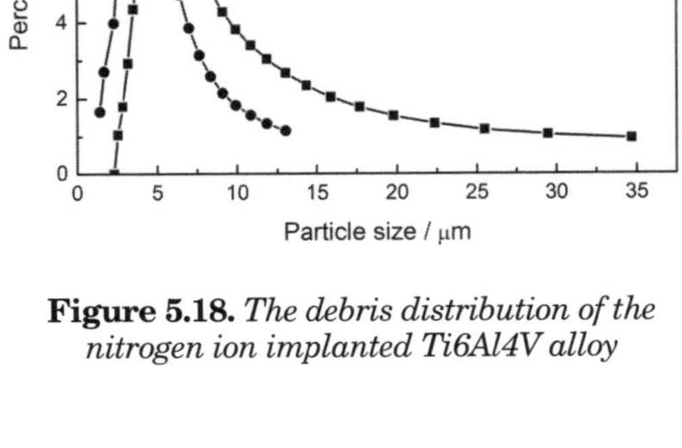

Figure 5.18. *The debris distribution of the nitrogen ion implanted Ti6Al4V alloy*

In summary, nitrogen ion implantation and carburization technology were chosen to modify the surface properties of titanium alloys. All of the results suggest that sequential carburization was an effective way to strengthen the titanium alloy surface and significantly improve the surface wettability and wear resistance.

5.3.3. *Artificial joint simulation*

A medical grade titanium alloy (Ti6Al4V) femoral head with the diameter of 28 mm was polished to Ra < 0.05 μm, then ultrasonically cleaned in ethanol and acetone, in turn, for 30 min, and inserted into a vacuum gas carburizing furnace. Prior to heating, the reaction chamber was first evacuated by a vacuum pump. Then a sequential carburization process was performed on the titanium alloy in the vacuum carburizing furnace and finally a titanium cermet femoral head was formed with a deep-black color and porous surface structure, as shown in Figure 5.19 [LUO 09d].

Figure 5.19. *Titanium cermet femoral head and UHMWPE cup*

A 1 million cycle *in vitro* hip simulator study between a titanium cermet femoral head and a UHMWPE cup was performed on an artificial hip joint simulator. A Paul cycle (ISO 14242-1 (2002)) with a peak of 784 N was applied with a 1 Hz cycle rate. 25% bovine serum with 0.1% sodium azide was chosen as a lubricant during the 1 million cycles, simulation and wear was determined gravimetrically after each 0.1 million cycles.

Figure 5.20 shows the stage mass loss of a UHMWPE cup counterfaced with a Ti6Al4V femoral head and a titanium cermet femoral head for one million cycles. We noted that the mass loss of the UHMWPE cup per 0.1 million cycles has a different tendency for the titanium femoral head and titanium cermet femoral head in the first million cycles.

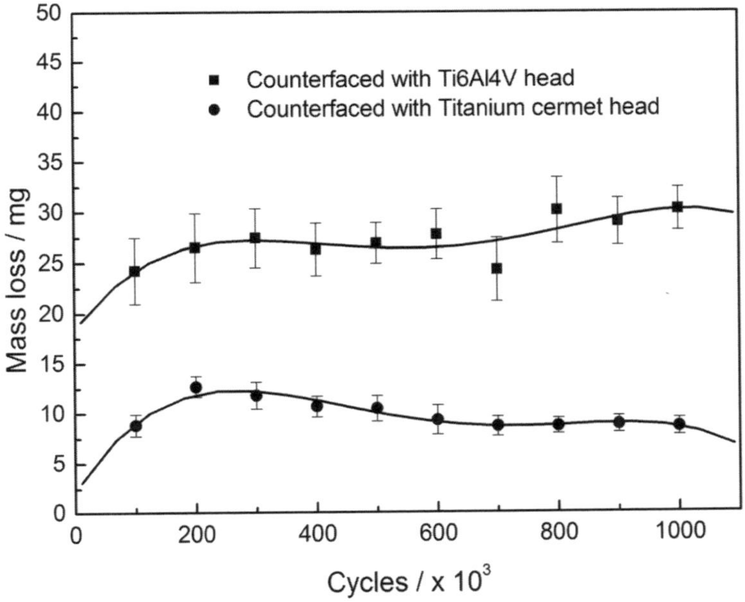

Figure 5.20. *Stage mass loss of UHMWPE cup counterfaced with a Ti6Al4V head and a carburized Ti6Al4V head*

The mass loss of UHMWPE cup per 0.1 million cycles keeps a steady increase when against the titanium femoral head, from 24.2 mg for the first 0.1 million cycles to 30.2 mg for the tenth 0.1 million. However, the mass loss of UHMWPE cup per 0.1 million cycles increases in the beginning and then keeps a decrease tendency when against titanium cermet femoral head.

In addition, titanium cermet femoral head can effectively reduce the wear of UHMWPE, with a decrease of 63.6% for the first 0.1 million cycles and 71.5% for the tenth 0.1 million cycles. As the sequential carburization increases the shear strength of titanium, and increases the surface hardness by forming a hard titanium carbide layer, it thus improves the friction and wear of the UHMWPE-Titanium cermet bearing.

Figure 5.21 shows the mass losses of the titanium femoral head and the UHMWPE cup after 1 million cycles. The titanium cermet femoral head has no mass loss after 1 million cycles and can significantly reduce the mass loss of UHMWPE cup compared to titanium femoral head, which decreases from 272.6 mg to 98.5 mg. Unlike the zero wear of the titanium cermet femoral head, some wear appears in the titanium femoral head, and the mass loss is 8.1 mg after 1 million cycles against a UHMWPE cup.

The wear of titanium will result in a lot of black debris, which will cause a reverse effect of cell response and lead to the failure of joints. Therefore, the wear of titanium proves that titanium alloy without any surface modification is not suitable for hip joint material.

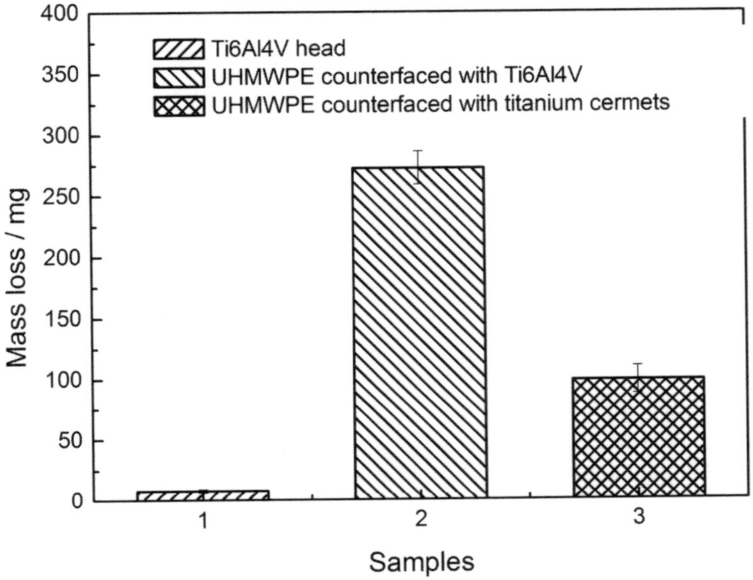

Figure 5.21. *Mass losses of three materials after one million cycles*

Figure 5.22 shows the debris distribution of UHMWPE against titanium and titanium cermet femoral heads. It is was found that the median diameters of UHMWPE wear debris when against titanium and titanium cermet femoral heads are 6.56 μm and 8.55 μm respectively.

In addition, the titanium cermet femoral head could effectively control the UHMWPE debris distribution, avoiding very large debris – more than 100 μm – thus helping the friction and wear of the bearing. The results indicate that titanium cermet femoral head can increase the size of UHMWPE wear debris and thus reduce the adverse effects on cellular response caused by wear debris, indicating potential biomaterials in biomedical fields.

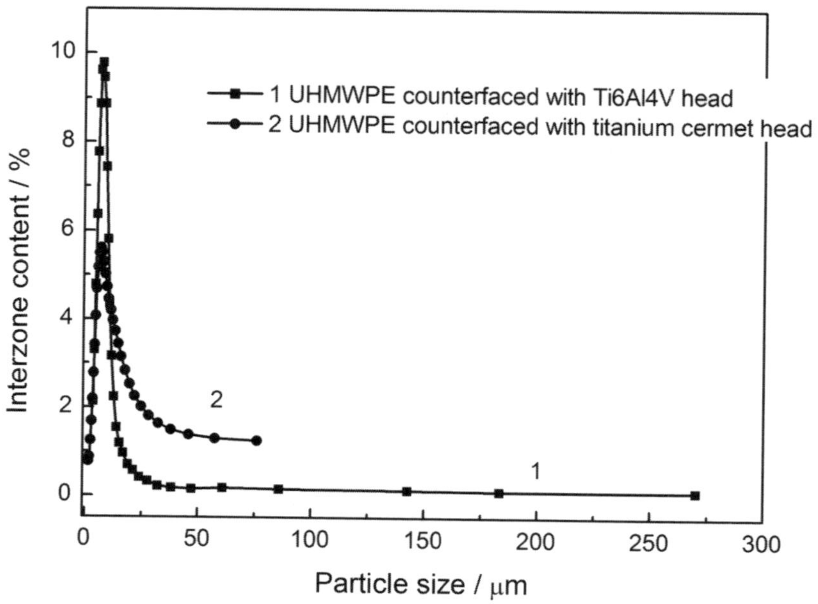

Figure 5.22. *The debris distribution of UHMWPE against titanium and titanium cermet femoral head*

In summary, the titanium cermet femoral head was successfully synthesized by using a novel sequential carburization under high temperature. The artificial hip joint experimental results show that the titanium cermet femoral head could not only improve the wear resistance of the femoral head, but also decrease the wear of the UHMWPE joint cup. In addition, the titanium cermet femoral head could effectively control the UHMWPE debris distribution, and increase the size of UHMWPE debris, thus decreasing the cell toxicity of debris.

All of the results suggest that the titanium cermet femoral head is a prospective femoral head material for artificial joints.

5.4. Acknowledgments

The author would like to acknowledge Elsevier, Springer Science and Professional Engineering Publishing for their kind permission to reuse text, tables and figures from previously published material [LUO 09a, LUO 09b, LUO 09c, LUO 09d, LUO 10].

5.5. Bibliography

[BEN 01] BENKO E., BARR T.L., HARDCASTLE S., HOPPE E., BERNASIK A., MORGIEL J., "XPS study of the cBN-TiC system", *Ceramics International*, vol. 27, p. 637-643, 2001.

[BRA 07] BRAMA M., RHODES N., HUNT J., RICCI A., TEGHIL R., MIGLIACCIO S., ROCCA C.D., LECCISOTTI S., LIOI A., SCANDURRA M., DE MARIA G., FERRO D., PU F.R., PANZINI G., POLITI L., SCANDURRA R., "Effect of titanium carbide coating on the osseointegration response in vitro and in vivo", *Biomaterials*, vol. 28, p. 595-608, 2007.

[DIC 90] DICASTRO V., POLZONETTI G., CONTINI G., COZZA C., PAPONETTI B., "XPS study of MnO2 minerals treated by bioleaching", *Surface and Interface Analysis*, vol. 16, p. 571-574, 1990.

[GOO 06] GOODMAN, S.B., MA, T., CHIU, R., RAMACHANDRAN, R., SMITH, R.L., "Effects of orthopedic wear particles on osteoprogenitor cells", *Biomaterials*, vol.27, p. 6096–6101, 2006.

[LEE 03] LEE D.W., KIM B.K., "Synthesis of nano-structured titanium carbide by Mg-thermal reduction", *Scr. Mater*, vol. 48, p. 1513-1518, 2003.

[LUO 09a] LUO Y., GE S., "Fretting wear behavior of nitrogen ion implanted titanium alloys in bovine serum lubrication", *Tribology International*, vol. 42, p. 1373-1379, 2009.

[LUO 09b] LUO Y., GE S., JIN Z., FISHER J., "Effect of surface modification on surface properties and tribological behaviours of titanium alloys", *Proceedings of the Institution of Mechanical Engineers Part J-Journal of Engineering Tribology*, vol. 223, p. 311-316, 2009.

[LUO 09c] LUO Y., GE S., JIN Z., "Wettability modification for biosurface of titanium alloy by the means of sequential carburization", *Journal of Bionic Engineering*, vol.6, p. 219-223, 2009.

[LUO 09d] LUO, Y., GE S., LIU H., JIN Z., "Microstructure analysis and wear behavior of titanium cermet femoral head with hard TiC layer". *Journal of Biomechanics*, vol.42, p. 2708-2711, 2009.

[LUO 10] LUO Y., GE S, JIN Z., FISHER J., "Formation of titanium carbide coating with micro-porous structure", *Applied Physics A*, vol. 98, p. 765-768, 2010.

[PAU 05] PAULIN C., FOUVRY S., DEYBER S., "Wear kinetics of Ti-6Al-4V under constant and variable fretting sliding conditions", *Wear*, vol. 259, p.292–299, 2005.

[SHI 04] SHI L., GU Y.L., CHEN L.H., YANG Z.H., MA J.H., QIAN Y.T. "Formation of nanocrystalline TiC by a low-temperature route", *Chemistry Letters*, vol. 33, p. 56-57, 2004.

[TAM 93] TAMADA Y., IKADA Y., "Cell adhesion to plasma-treated polymer surfaces", *Polymer*, vol.34, p. 2208–2212, 1993.

[VAD 06] VADIRAJ A., KAMARAJ M., "Characterization of fretting fatigue damage of PVD TiN coated biomedical titanium alloys", *Surface & Coatings Technology*, vol. 200, p. 4538–4542, 2006.

[VAD 07] VADIRAJ A., KAMARAJ M., "Effect of surface treatments on fretting fatigue damage of biomedical titanium alloys", *Tribology International*, vol.40, 82–88, 2007.

[WAN 02] WANG D.Y., CHIU M.C., "Characterization of TiN coatings post-treated by metal–plasma ion implantation process", *Surface & Coatings Technology*, vol. 156, p. 201–207, 2002.

[XIA 98] XIAO S.Q., LUNGU C.P., TAKAI O., "Comparison of TiN deposition by rf magnetron sputtering and electron beam sustained arc ion plating", *Thin Solid Films*, vol. 334, p.173–177, 1998.

[XIN 98] XIN H., WATSON L.M., BAKER T.N., "Surface analytical studies of a laser nitrided Ti- 6Al-4V alloy: a comparison of spinning and stationary laser beam modes", *Acta Materialia*, vol. 46, p. 1949–1961, 1998.

[YIN 05] YIN X., GOTMAN I., KLINGER L., GUTMANAS E.Y., "Formation of titanium carbide on graphite via powder immersion reaction assisted coating", *Mater. Sci. Eng. A*, vol. 396, p. 107-114, 2005.

List of Authors

Claire BROCKETT
Institute of Medical and Biological Engineering
University of Leeds
UK

J. Paulo DAVIM
Department of Mechanical Engineering
University of Aveiro
Portugal

John FISHER
Institute of Medical and Biological Engineering
University of Leeds
UK

Mathias Christian GALETZ
Dechema e.V.
Karl-Winnacker-Institut
Frankfurt-am-Main
Germany

Uwe GLATZEL
University of Bayreuth
Bavaria
Germany

J. Philippe KRETZER
Laboratory of Biomechanics and Implant Research
Department of Orthopedics Traumatology and Paraplegiology
Heidelberg University Hospital
Germany

Yong LUO
Institute of Tribology and Reliability Engineering
School of Material Science and Engineering
China University of Mining & Technology
Xuzhou
China

Vesa SAIKKO
Department of Engineering Design and Production
Aalto University School of Science and Technology
Finland

Index

T, U, W